APPLYING THE FLOOD VULNERABILITY INDEX AS A KNOWLEDGE BASE FOR FLOOD RISK ASSESSMENT

Applying the Flood Vulnerability Index as a Knowledge base for flood risk assessment

DISSERTATION

Submitted in fulfillment of the requirements of

the Board for Doctorates of Delft University of Technology

and of the Academic Board of the UNESCO-IHE

Institute for Water Education

for the Degree of DOCTOR

to be defended in public on

Wednesday, 6th of June 2012, at 10:00 hours

in Delft, the Netherlands

by

Stefania - Florina BALICA

Master of Science in Hydraulic Engineering and River Basin Management
UNESCO-IHE, Delft, the Netherlands

born in Drobeta Turnu Severin, România.

This dissertation has been approved by the supervisor(s):
Prof. Dr. Nigel G. Wright
Prof. Dr. Ir. Arthur E. Mynett

Composition of Doctoral Committee:

Chairman	Rector Magnificus TU Delft
Vice-Chairman	Rector UNESCO-IHE
Prof. Dr. N.G. Wright	UNESCO-IHE, Delft University of Technology, supervisor
Prof. Dr. Ir. A.E. Mynett	UNESCO-IHE, Delft University of Technology, supervisor
Prof. Dr. H. Guangwei	Sophia University, Japan
Prof. Dr. P. Gourbesville	Nice-Sophia Antipolis University, France
Prof. Dr. Ir. M.J.F. Stive	Delft University of Technology
Prof. Dr. Ir. W.A.H. Thissen	Delft University of Technology
Prof. Dr. Ir. N.C. van de Giesen	Delft University of Technology, reserve member

CRC Press/Balkema is an imprint of the Taylor & Francis Group, an informa business

Published by:
CRC Press/Balkema
PO Box 447, 2300 AK Leiden, the Netherlands
e-mail: Pub.NL@taylorandfrancis.com
www.crcpress.com - www.taylorandfrancis.co.uk - www.ba.balkema.nl

ISBN 978-0-415-64157-9 (Taylor & Francis Group)

" *Nosce te ipsum* " - *Know Thyself, inscribed in the pronaos (forecourt) of the Temple of Apollo at Delphi according to the Greek periegetic (travelogue) writer Pausanias (2nd century AD)*

To vulnerable systems

Content

Nomenclature and symbols

The following nomenclature and symbols were used in this disseration.

Abbrevations

ADB – Asian Development Bank
AIACC - Assessments of Impacts and Adaptations to Climate Change
ASCE – American Society of Civil Engineers
CCFVI – Coastal Cities Flood Vulnerability Index
CRED – Centre for Research on the Epidemiology of Disasters
CSoVI – Coastal Social Vulnerability Index
CVI – Climate Vulnerability Index
DEWA – Division of Early Warning and Assessment
DFID – Department for International development of the United Kingdom
DRI – Disaster Reduction Institute
E - Exposure
EC-JRC – European Commission – Joint Research Centre
EUROSION – A European initiative for sustainable coastal erosion management
FVI - Flood Vulnerability Index
FVIEc – Economic component of FVI
FVIEn – Environmental component of FVI
FVIHydro-Geological – Hydro-Geological component of CCFVI
FVIPh - Physical component of FVI
FVIS – Social component of FVI
GIS – Geographical Informational System
GRAVITY – Global Risk and Vulnerability Trends per Year
ICHARM – International Centre for Water Hazard and Risk Management under the auspices of UNESCO
IPCC – Intergovernmental Panel on Climate Change
ISDR – International Strategy for Disaster Reduction
IUCN - International Union for Conservation of Nature
IWMI – Integrated Water Management Institute
MLIT -Ministry of Land, Industry and Transport
Munich Re - Reinsurance, primary insurance and Munich Health
NS – Natural Subsystem
OECD – Organisation for Economic Co-operation and Development
PAC – Politico-Administrative Subsystem
PAGASA - Philippines Atmospheric, Geophysical and Astronomical Services Administration
PCA – Principal Component Analysis
PHP – Hypertext Preprocessor
PREVIEW – Geneva's Project for Risk Evaluation, Vulnerability, Information and Early Warning
R - Resilience
RB – River Basin
RWI – Risk Watch International
S - Susceptibility
SC – Sub-Catchment
SEA – Strategic Environmental Assessment
SES – Socio-Economic Subsystem
SFVI – Social Flood Vulnerability Index
SRES – Special Report on Emissions Scenarios
SVCC – Social Vulnerability to Climate Change for Africa

SVI – Social Vulnerability Index
UN – United Nations
UNEP – United Nations Environmental Programme
UNESCAP – United Nations Economic and Social Commission for Asia and the Pacific
UNESCO – United Nations Educational, Scientific, and Cultural Organization
UNICEF – United Nations Children's Fund
WB – World Bank
WPI – Water Poverty Index

Symbols
cyc – number of cyclones
% disable – percent of disable people
A/P – Awareness/Preparedness
AmInv – Amount of Investments
CH – cultural heritage
CL – coastal line
C_m – Child Mortality
C_{PR} – Communication Penetration Rate
D_L – Dikes & Levees
D_A – Degraded Area
E_{CR} – Economic Recovery
E_R – Evacuation Roads
E_V - Evaporation
FHM – Flood Hazard Maps
F_i – Flood Insurance
FP – Flood Protection
FS – Foreshore Slope
GCP – growing coastal population
GDP – Gross Domestic Product
HDI – Human Development Index
I_{neq} - Inequality
IO – Institutional Organizations
L_{EI} – Life Expectancy Index
L_U - Land Use
N_R – Natural reservations
PCL – population close to coastal line
P_E – Past Experience
P_{FA} - Population in flood prone area
$R_{ainfall}$ - Rainfall
RD – river discharge
R_{Pop} – Rural Population
RT – recovery time
S – Shelters
S_C – Storage Capacity
SLR – sea level rise
SS – Storm Surge
T - Topography
U – Urban Area
U_A – Urbanised Area
U_G – Urban Growth
U_M - Unemployment

U_{npop} – Unpopulated Areas
UP – Uncontrolled Planning Zone
V_{year} - Volume yearly
W_s – Warning System

Summary

Floods are one of the most common and widely distributed natural risks to life and property worldwide. There is a need to identify the risk of flooding in flood prone areas to support decisions for flood management from high level planning proposals to detailed design.

An important part of modern flood risk management is to evaluate vulnerability to floods. *This evaluation can be done only by using a parametric approach.* Worldwide there is a need to enhance our understanding of vulnerability and to also develop methodologies and tools to assess vulnerability. One of the most important goals of assessing flood vulnerability is to create a readily understandable link between the theoretical concepts of flood vulnerability and the day-to-day decision-making process and to encapsulate this link in an easily accessible tool. Therefore, the main objective of this thesis is to make this assessment more straightforward while demonstrating the applicability of the improved FVI methodology (a parametric approach), so it can contribute to the development of the existing knowledge based on flood risk assessment methodologies.
The main contributions of this Ph.D. research are:

- definition and improvement of the FVI methodology;
- its practical application to different spatial scales;
- climate change consideration for FVI indicators and FVI forecast for coastal cities;
- comparison with the deterministic methods for flood risk assessment;
- creation of a collaborative tool for FVI transfer of knowledge and results.

The research starts by portraying a methodology using indicators to calculate a FVI which is intended for assessing the conditions which influence flood damage at various spatial scales: river basin, sub-catchment, urban area and coastal flood damage at city scale.

The methodology developed distinguishes different characteristics at each identified spatial scale, thus allowing a more in-depth analysis and interpretation of local indicators. An indicator, or set of indicators, can be defined as an inherent characteristic which quantitatively estimates the condition of a system; indicators are simple numbers which encapsulate knowledge of the system, for example: the rate of mortality in a region, the GDP per capita, the storage capacity of a dam. These indicators embrace general information on age, poverty, gender, race, education, social relations, institutional development, and population with special needs. Understanding each concept and considering certain indicators may help to characterise the vulnerability of different systems, by which certain actions can be identified to decrease it. *The parametric approach, here the FVI, through indicators is the only one which evaluates vulnerability to natural hazards, such us floods;* This also pinpoints local hotspots of flood vulnerability.

The final results are presented by means of a standardised number, ranging from 0 to 1, which symbolises comparatively low or high flood vulnerability between the various spatial scales.

Next, for the FVI methodology to be sustainable, *improvements* were made by analysing the indicators' relevance and by studying the main indicators needed to portray the reality of flood vulnerability in an effective way. For this purpose, mathematical tools (a derivative and a correlation method) and expert knowledge (via a questionnaire) were used. Finally, all these methods were combined in order to select the most significant indicators and to simplify the FVI equations. After reducing its complexity, the FVI can be more easily used as a tool for education, improvement of decision making and ultimately reduction of flood risk.

The same flood vulnerability approach was implemented for coastal cities, in order to study the impact of climate change on the vulnerability of these cities over a longer timescale than the present. The results show that coastal cities flood vulnerability index can provide a means of obtaining a broad overview of flood vulnerability and the effect of possible adaptation options. This, in turn, can allow for the direction of resources to more in-depth investigation of the most promising strategies.

In order to compare and justify the FVI methodology, the traditional and commonly used methods of computer modelling and inundation mapping are examined. The research focuses on the applicability and performance of the FVI method for flood risk assessment in comparison to the traditional methods. This is undertaken in a data scarce area (Budalangi area, Kenya, using the SOBEK 1D/2D model). The methods gave comparable information in data scarce areas. However the research indicated that the FVI methodology is more appropriate for high level decision making while traditional modelling methods are more appropriate for the design stage. The conclusions of this comparison indicate that a combination of methods should be used in flood risk planning and assessment.

An automated calculation of a flood vulnerability index implemented through a web management interface that enhances the ability of decision makers to strategically guide investment is presented (http://www.unesco-ihe-fvi.org). A network of knowledge between different institutions and universities in which this methodology is used was created. It is also has encouraged collaboration between the members of the network on managing flood vulnerability information and also promoting further studies on flood risk assessment at all scales.

In conclusion, this thesis provides a holistic approach to be used in flood vulnerability assessment and in this way hopes to facilitate the consideration of system impacts in water resources decision-making. This approach was verified in practical applications on different spatial scales and comparison with traditional methods. The FVI developed in this thesis shows that the FVI tool can be applied in a broad range of contexts (river and coastal floods, including their conditions of the applied index components and indicators). As well this methodology can produce helpful understanding into vulnerability and capacities for using it in planning and implementing projects. *The FVI was developed as a unique approach to evaluate vulnerability and based on it to prioritise investments and to respond to a flood disaster by understanding what impact future actions will have on vulnerabilities in place.*

Samenvatting

Overstromingen behoren tot de meest voorkomende natuurrampen die wereldwijd levensbedreigende en verwoestende gevolgen kunnen hebben. Er bestaat dan ook een dringende behoefte om deze overstromingsrisico's te kunnen bepalen teneinde maatregelen te kunnen voorbereiden, variërend van voorstellen op het gebied van ruimtelijke ordening tot het gedetailleerd ontwerpen van veiligheidsmaatregelen.

Een belangrijke component om overstromingsrisico's te beperken is om de kwetsbaarheid tegen overstromingen vast te stellen. *Deze kan alleen worden verkregen door een parametrische aanpak te volgen.* Over de hele wereld bestaat de noodzaak om een beter begrip te krijgen van de kwetsbaarheid tegen overstromingen. Vandaar dat methoden en technieken worden ontwikkeld om deze kwetsbaarheid te bepalen. Een belangrijk onderdeel daarvan is om een beter verband te leggen tussen theoretische concepten van overstromingsrisico's en de dagelijkse praktijk van besluitvorming – en om deze link in een gemakkelijk toegankelijk instrument te vatten.

Het hoofddoel van dit proefschrift is dan ook om dit proces te ondersteunen door het ontwikkelen van een verbeterde "Flood Vulnerability Index (FVI)" methodologie (een parametrische aanpak), die kan bijdragen aan de verdere ontwikkeling van de bestaande kennisbasis om overstromingsrisico's vast te stellen.

De belangrijkste bijdragen van dit promotie onderzoek zijn
- het beschrijven en verbeteren van de FVI methodologie;
- het laten zien van praktische toepassingen op verschillende ruimteschalen;
- het meenemen van effecten van klimaatverandering op FVI indicatoren en FVI voorspellingen voor steden in kustgebieden;
- het vergelijken met deterministische methoden voor het bepalen van overstromingsrisico's
- het ontwikkelen van een instrumentarium voor samenwerking en kennisoverdracht op het gebied van FVI.

De eerste fase van het onderzoek is gericht op het ontwikkelen van een methodologie waarbij gebruik wordt gemaakt van indicatoren om een FVI index te bepalen die de schade door overstromingen op verschillende ruimtelijke schalen kan weergeven: van integrale stroomgebieden tot deel-stroomgebieden tot stedelijke conglomeraties in kustgebieden. De ontwikkelde methodologie bepaalt de verschillende karakteristieken op elk van deze schalen waardoor een meer diepgaande analyse van elk van de lokale indicatoren mogelijk wordt.

Een indicator, of groep van indicatoren, kan worden gekarakteriseerd als een grootheid die de toestand van een systeem in kwantitatieve zin bepaalt; de indicatoren zelf zijn eenvoudige getalsmatige representaties van het systeem, bijv. de sterftecijfers in een bepaalde regio, het inkomen per hoofd, of de bergingscapaciteit van een reservoir. Deze indicatoren omvatten algemene informatie over leeftijdsopbouw, inkomensverdeling, bevolkingssamenstelling, onderwijsniveau, sociale relaties, bestuurlijke structuren, en de aanwezigheid van bevolkingsgroepen met speciale behoeften.

Kennis van deze indicatoren kan worden gecombineerd teneinde de kwetsbaarheid van bepaalde gebieden te bepalen zodat maatregelen kunnen worden getroffen. *Een dergelijke parametrische aanpak met behulp van indicatoren, in dit geval FVI, is de enige manier om de kwetsbaarheid van bepaalde gebieden vast te stellen tegen natuurrampen, zoals overstromingen.* Bovendien kunnen op deze manier gevarenzones worden gelokaliseerd die erg kwetsbaar zijn tegen overstromingen. Het uiteindelijke resultaat wordt gepresenteerd door middel van een genormeerd getal variërend tussen 0 en 1, overeenkomend met een lage dan wel hoge kwetsbaarheid van het gebied.

Vervolgens zijn *verbeteringen* aangebracht in de FVI methodologie door de relevantie van de verschillende indicatoren na te gaan en de belangrijkste componenten te bepalen die de kwetsbaarheid tegen overstromingen het best weergeeft. Daartoe zijn wiskundige technieken gebruikt (een gradiënten methoden en een correlatie techniek) tezamen met de expertise van specialisten (op basis van vragenlijsten). Uiteindelijk werden al deze methoden gecombineerd teneinde de meest belangrijke indicatoren te bepalen en de FVI vergelijkingen te vereenvoudigen. Op deze manier wordt de FVI beter hanteerbaar bij kennisoverdracht, besluitvorming en het beperken van overstromingsrisico's.

Eenzelfde benadering werd toegepast op steden in kustgebieden, teneinde het effect van zeespiegelrijzing ten gevolge van klimaatverandering te onderzoeken op langere tijdschaal. De resultaten laten zien dat FVI voor steden in kustgebieden direct een beeld geeft van de kwetsbaarheid en van mogelijke tegenmaatregelen. Daarmee bestaat de mogelijkheid om gericht te zoeken naar de meest veelbelovende strategie om de kwetsbaarheid tegen overstromingen te verminderen.

Teneinde de FVI benadering te kunnen vergelijken en beoordelen, zijn de meer traditionele en vaak gebruikte methoden op basis van computermodellen en overstromingskaarten bestudeerd. Daartoe is een casus in een gebied met beperkt beschikbare gegevens gebruikt (Budalangi in Kenya, op basis van het SOBEK 1D/2D model). Beide methoden gaven vergelijkbare uitkomsten, waarbij de FVI methodologie met name geschikt bleek voor besluitvorming op hoog niveau, terwijl de traditionelere methoden meer geschikt waren voor detailontwerp. Een combinatie van beide methoden is aan te bevelen bij het beoordelen van overstromingsrisico's en het ontwikkelen van tegenmaatregelen.

Een geautomatiseerde berekeningsmethode voor het bepalen van FVI via een web management interface is beschikbaar op (http://www.unesco-ihe-fvi.org); hiermee kunnen beleidsmakers strategische beslissingen voorbereiden en investeringsvoorstellen afwegen. Ook is een *kennisnetwerk* opgebouwd tussen diverse instellingen en universiteiten. Dit heeft de samenwerking bevorderd tussen de verschillende partners over hoe overstromingsrisico's in te schatten en tegenmaatregelen te bevorderen.

Samenvattend kan worden gesteld dat dit proefschrift een holistische methodiek verschaft voor het vaststellen van overstromingsrisico's en het treffen van tegenmaatregelen. De methodiek is beproefd in praktische toepassingen op verschillende ruimtelijke schalen en vergeleken met meer traditionele methoden. De FVI benadering als ontwikkeld in dit proefschrift laat zien dat de FVI benadering kan worden ingezet bij een groot scala aan toepassingen (overstromingen van zowel rivieren als kustgebieden). Bovendien kan de FVI aanpak bijdragen om tegenmaatregelen te ontwikkelen die vervolgens in projecten kunnen worden uitgevoerd. *De FVI benadering is een unieke aanpak om de kwetsbaarheid tegen overstromingen na te gaan en benodigde investeringen af te wegen die gemoeid zijn met het nemen van tegenmaatregelen die pas op termijn tot resultaten zullen leiden.*

CHAPTER 1

Introduction

1.1. What are floods?

Lately, the frequency of floods and flooding is increasing; many flood events have now been studied by many authors (Ouarda et al., 200;, Gaume, et al., 2010; Villarini, et al., 2009; Schmocker-Fackel & Naef, 2010; Greenbaum et al., 2010; Prudhomme et al., 2003; Kusumastuti et al., 2008).

For this thesis a distinction has to be made between the two terms, floods and flooding, which are frequently confounded, when subjects relating to high water stage or crest discharge are considered. Below are defining the terms as:

A flood is defined as "a temporary condition of surface water (river, lake, sea), in which the water level and/or discharge exceeds a certain value, thereby escaping from its normal confines"; this does not necessarily result in flooding (Douben, 2006a, Schultz, 2006).

Flooding is defined as "the spilling over or failing of the normal limits for example steam, lake, sea or accumulation of water as a result of heavy precipitation through lack or beyond of the discharge capacity of drains, or snow melt, dams or dikes break affecting areas" (Douben and Ratnayake, 2005), which are normally not submerged (Ward, 1978).

1.1.1. Types of floods

A distinction can be made between four different types of floods: coastal floods, river floods, flash floods and urban floods (MunichRe, 2007).

1.1.1.1.   Coastal floods

These can happen all along the coast and also alongside banks of large lakes (MunichRe, 2007). Floods usually occur when storms coincide with high tides and can include overtopping or breaching of beaches. Coastal flooding may also happen by sea waves called tsunamis, unusually huge tidal waves due to volcano or earthquake activity in the ocean. Tropical storms and hurricanes can generate serious rains, or drive ocean water into land. These floods can create the potential for extreme loss and may cause a large number of casualties.

The accelerating rise in sea levels that is undoubtedly to be anticipated as an effect of climate change and variability will intensify the risk of storm surges and coastal erosion around the world — the sea level rise will be considered one of the most harmful effects of global warming; Coastal flooding levels (NYC Hazards, 2007) — classified as minor, moderate or major — are computed based on the quantity of water that rises above the usual tide in a particular area. Coastal flooding can be very destructive (Natural Environmental, 2007).

Coastal and estuarine floods take place when the sea level rises further than its normal fluctuations or/and in conjunction with high river flows; land subsidence and progressive sea level rise are also aspects that raise the height of the sea level more than its normal fluctuations. Coastal areas and low-lying island states are vulnerable to this sort of flooding.

1.1.1.2.   River floods

Floods along rivers are a natural event. River floods occur when the spring rains and with winter snows melt merge. The river basins are filled too fast, and then the stream will spill over its banks. River floods can also occur because of heavy rainfall for a period of days over a large area. The soil becomes drenched and cannot cope with any more water so that the rain flows directly into the rivers (Hoyt, 1955).

River floods do not occur suddenly but develop gradually – even if in a short time. River floods can last from a few days to a few weeks. The flooded area could be very wide if the river valley is exposed and large and the river carries a huge volume of water. River related flooding also brings indirect threats arising from food and drinking water shortage and the spreading of diseases (Douben, 2006b).

1.1.1.3.   Flash floods

Flash floods are temporary inundations of different areas such as: river basins, sub-catchments and a town or parts of a city. Short periods of intense rain can cause flash floods, they usually occur in combination with thunderstorms and over a very small area. The ground is not usually soaked; but at the rainfall intensity exceeds the infiltration rate, the water runs off the surface and soon collects in the receiving waters.

Flash floods can take place almost anywhere, so that almost everybody is vulnerable. Sometimes, flash floods, predict the beginning of a major river flood, but usually they are split, individual events of only local significance, scattered randomly in space and time. Engineered works, as dams, dikes and levees, are put in place for flood protection. They usually are constructed to endure an inundation with a computed risk of occurrence. A dam, dike or levee may be calculated to hold a flood at a specific location on a river that has a certain probability of occurring. If a bigger flood occurs, then that construction will theoretically be overtopped. While overtopping the structure will fail or washed out, this water will to become a flash flood (Perry, 2000, Kron, 2005).

Flash floods kill and damage the most. These kinds of floods occur without warning and transport huge amounts of fast-moving water. Unfortunately, they are also the most widespread sort of flood. Regarding the time, flash floods, happen in shorter time than river floods. The main water capacity will run off again after a few hours.

1.1.1.4.   Urban floods

Urban floods are usually caused by extreme local rainfall, combined with blocked drainage systems. This type of flooding depends on soil and topographical conditions and the quality of the drainage system (Douben, 2006b).

These urban floods, are increasing, they are the effect of urban/suburban sprawl, where urbanized land is not capable of rainfall absorption.

Urban floods occur mostly as a result of the impermeability of buildings and roads. In time of heavy precipitation, the large amount of rain water cannot be absorbed into the ground and leads to urban runoff. The urban floods depend on the topography and soil conditions (Douben, 2006). These types of flood, with different flow regimes and interventions strategies can create different kinds of vulnerability in river basins, sub-catchments and urban areas, depending on the local situation.

1.1.2. Frequency of floods

Floods are natural and recurring events in a river or stream. Floods are commonly described in terms of their statistical frequency. A 100-year flood or 100-year floodplain describes a flood or a place subject to a 1% probability of that flood to occur in a given year.

It is well-known that the frequency of flood is based on the environment, the river bank material, and river slope. In regions with heavy rainfall each year, or where the annual flood is derived from snow melts, the floodplain may be inundated almost each year, even along large rivers with very small channel slopes. In regions with high temperatures, floods generally occur in the period of highest precipitation (United States Agency, 1991). In some areas floods occur because of exposure to the cyclones, hurricanes, big tidal waves or tsunamis.

Lately, any type of floods raise subject matter of concern for people, authorities, insurance companies and policy makers; for that reason an increased focus on flood management and flood mitigation is urgent.

1.2. Flood management and flood mitigation

Many studies describe the possible causes and effects of floods in terms of loss of human lives and costly damages and possible counter measures that can be adopted to minimize their consequences (Hall et al., 2003; Sayers et al., 2002; Connor & Hiroki, 2005; Naess et al., 2005, Nicholls, 2004; Plate, 2002; Montz & Gruntfest, 2002; Mustafa, 2003).

Flood risk management has various approaches to reduce floods (to some extent) and for mitigating their consequences. Flood management is an extensive range of water resources activities aimed at reducing possible destructive impacts of floods on citizens, environment and economy of a area.

Flood risk management is defined as all activities that aim at sustaining or improving the capability of a region to cope with floods. Risk is defined as a function of flood probabilities and flood impacts.

Some objectives of flood risk management are specific final results that have to be accomplished in a prearranged time frame. These are:

- diminishing exposure of citizens, property and environment to flood risks
- diminishing the present level of flood damage
- increasing the resilience of people and systems.

Roughly, the approaches for flood mitigation and defence can be divided into two: structural and non-structural measures. The aim of structural measures is to modify the flood pattern, while non-structural measures aim at reduction of the flood impacts (Parker, 2000).

1.3.1. Structural measures

The structural measures consist of infrastructure development that modifies the river flow, like dams, barrages, dikes, levees, channeling, etc. that reduce floods from causing damages to the population or infrastructure in the flood prone area (Douben, 2006b). The basic principles consist of storing, diverting and/or confinement of floods. They usually consist of large investments for large engineering structures, which sometimes are inevitable to preserve the safety and development of a region.

1.3.2. Non-structural measures

Mitigation measures rely on the flood observation and ability of people to prepare where disaster happens. Non-structural measures consist of several mitigation measures not modifying the river flow; such as: planning, programming, setting policies, co-coordinating, facilitating, raising awareness, assisting, preparedness, response, legislation, flood forecasting and warning systems, flood proofing, flood fighting, post-flood rehabilitation financing, reconstruction and rehabilitation planning (Andjelkovic, 2001), as well as insuring, educating, training, regulating, reporting, informing and assessing. "If structural measures are the metallic frames of a flood management program, then mitigation is its cloak" (Andjelkovic, 2001). Mitigation measures are traditionally referred to as non-structural measures.

It is very important to know in which order to apply these mitigation measures: primary, to develop public awareness and the political will, laws and regulations, secondly, risk reducing measures, and finally, education and organise out training. Finally a flood insurance industry (Kron, 2008) "should be created in order to potentially spread high flood damage cost over a long period of time and among a large number of people", (therefore the actions of reducing social, economic, environmental and physical vulnerability). These actions can be undertaken at river basin, sub-catchment, urban and even district level.

The evolution of non-structural measures is also linked with the need to improve the decision making process for flood protection, so that investments can be allocated in a more optimal way. For this purpose the introduction of indices for flood vulnerability, or other related issues is need and can be helpful. Non-structural measures and their component techniques, contribute directly towards reducing losses of life and damage to property.

A way of reducing losses of life and damages is to evaluate vulnerability to floods, defined as the extent to which a system is susceptible (McCarthy et al., 2001, Rao, 2005) to floods due to exposure, in conjunction with its ability (or inability) to cope, recover, or basically adapt.

An assessment of flood vulnerability is the Flood Vulnerability Index (FVI), a non-structural measure, more in the interest of general public, policy, decision makers and reinsuring companies. The index can be incorporate into the Disaster Risk Reduction Strategy. The FVI is assessed in a comprehensive way, taking into account all factors that are most likely to be affected by a flood disaster: social, economic/financial, environmental and physical aspects. The FVI, by aggregating indicators, can be used to assess vulnerability level of each vulnerability factor.

1.3. Research approach

1.3.1. Motivation and scope of the thesis

In 2005, Connor & Hiroki, developed the Flood Vulnerability Index (FVI) for river basins. Their methodology assesses flood vulnerability on a river basin scale by identifying different components that influence the susceptibility to floods of the people who live in flood prone areas. Connor and Hiroki's FVI identifies four main components; *climate, hydro-geological, socio-economic* and *existing counter measures*, derived from eleven indicators. The index analyses two aspects; the human index (FVI_H) which takes into account the loss of life, and the economic index (FVI_M) which considers the material losses caused by flooding events.

To further develop the methodology and calculate the FVI there is a need for smaller spatial scales sub-catchments, river and coastal urban areas. One of the problems encountered relates to the homogeneity of large areas, which can lead to unrealistic results, involving relatively high investments for monitoring and evaluating the necessary data. Another problem pictures the elusion of some indicators which may reflect a higher or lower vulnerability to floods. An improved FVI, seeking a more transparent communication process, i.e. vulnerability factors, to general public, political committees, analysts, investors and science is needed.

The most important goal of assessing flood vulnerability, in particular, is to create a readily understandable link between the theoretical concepts of flood vulnerability and the day-to-day decision-making process and to encapsulate this link in an easily accessible tool, which aims to identify hotspots related to flood events in different regions of the world. The goal is converting knowledge into actions: to assess/index the flood vulnerability in different regions of the world, but also in less well-served ones, data scarce areas, by following a uniform approach; raise flood vulnerability awareness; save lives, reduce economic, environmental losses and better distribute the financial burden.

An aim of the FVI is to be used as a toolkit to assess and manage the flood vulnerability and in this way to facilitate adaptation and coping capacities. By knowing system's vulnerability, should be known how to adapt and cope with floods, should be known how prepared and aware the systems are; This way resilience should be build vis-à-vis weather-related vulnerabilities.

Therefore the thesis research objectives are:

1.3.2. Research Objectives

General Objective: To demonstrate the applicability of the improved FVI methodology, so it can contribute to the development of the existing knowledge based on flood risk assessment methodologies.

Specific Objectives:

> ➢ To develop and apply the FVI methodology to various spatial scales, such as: river basin, sub-catchment and urban area.

> ➤ To analyse and reduce the complexity of the FVI by using varying mathematical and statistical methods.
> ➤ To develop and apply the FVI for coastal cities based on existing approach.
> ➤ To investigate the possibility of comparing the FVI methodology versus traditional flood modelling techniques.
> ➤ To gain academic acceptance by creating a network of knowledge in different countries with different socio-economic situations, based on the FVI.

1.3.3. Outline of the Thesis

The present dissertation comprises eight chapters, as follows (see **Error! Reference source not found.**):

- The current chapter (Chapter No. 1) introduces the context of the research, background and objectives.

- In Chapter 2 a brief literature review is presented concerning the definition of vulnerability, vulnerability to floods, vulnerability factors, indicators, also various indices and flood risk expressions.

- In Chapter 3, development and applications of the FVI methodology for three different spatial scales are provided. In this chapter, a downscaling analysis of various spatial scales is presented. As well Chapter 3 presents an automated calculation of a flood vulnerability index implemented through a web management interface (PHP). Here many case studies were required in order to cover the full range of cases in terms of scale such as river basin, sub-catchment and urban area.

- In order to have user friendly and less time consuming FVI methodology, Chapter 4 analyses various mathematical, statistical and surveying methods to simplify the FVI methodology by reducing the number of used methodology indicators, from the initial 71 indicators only 28 were retained.

- Since most of the research concerns the development and the application of the methodology, Chapter 5 presents the possibility to investigate the comparison of a parametric approach, the FVI methodology, and a deterministic approach, flood modelling, such us: SOBEK.

- Chapter 6 focuses on developing a Coastal City Flood Vulnerability Index (CCFVI) based on exposure, susceptibility and resilience to coastal flooding. It is applied to nine cities around the world, each with different kinds of exposure.

- In the last chapter (Chapter 7), overall conclusions, recommended strategies for further FVI applications, as well as comments and ideas for further works, are discussed.

CHAPTER 2

Synthesising vulnerability and risk

Parts of this chapter has been published as:

Balica, S.F., 2012, Approaches of understanding vulnerability indices developments to natural disasters, Environmental Engineering and Management Journal, June, 2012, Vol. 12 (6).

2.1. Introduction

As Chapter 1 explained, this thesis mainly aims to demonstrate the applicability of the improved FVI methodology, so it can contribute to the development of the existing knowledge based on flood risk assessment methodologies.

This second chapter explores the concept of vulnerability in order to understand the flood vulnerability and to validate/compare a vulnerability index, what are the indicators and the existing vulnerability indices, how the defined factors of vulnerability are being addressed in the approaches used.

Section 2.2 describes the concept of vulnerability and the concept of flood vulnerability and the flood vulnerability indices; Section 2.3 discusses the perception of flood vulnerability; the development and application of the flood vulnerability index will be also discussed in Chapter 3. Section 2.4 discusses the validity of vulnerability indices; flood risk expressions. Section 2.5 describes the flood risk expressions and Section 2.6 focuses on uncertainty in flood vulnerability and how FVI can be an indirect tool in reducing it.

2.2. Conceptualizing vulnerability – Who and what is vulnerability

While the concept of vulnerability is frequently used within disaster research, researchers' notion of vulnerability has changed over the past two decades and consequently there have been several attempts to define and capture what is meant by the term. After McEntire, 2010, Hufschmidt, 2011, analyse and compare a wide number of vulnerability concepts to natural hazards, throughout the view of different schools.

In the variety of definitions to vulnerability, the definition of hazards exposed to societies differs. Various definitions of vulnerability refer to climate change (IPCC, 1992, 1996 and 2001), others to environmental hazards (Blaikie et al., 1994); (Klein and Nicholls, 1999), (ISDR, 2004), and are several definitions of vulnerability to floods (Veen & Logtmeijer 2005, Connor & Hiroki, 2005, Balica et al., 2009, UNDRO, 1982, McCarthy et al., 2001).

By now it is generally understood that "vulnerability is the root cause of disasters" (Lewis, 1999, and later Wisner et al., 2004) and "vulnerability is the risk context" (Gabor and Griffith, 1980). The definition of Timmerman in 1981 includes the degree of a harmed system at risk, the frequency of a hazardous incident. The quality and the degree of the feedback are conditioned by the system's resilience. Chambers (1989), described vulnerability "as a potential for loss", with two sides: the shocks and perturbations from outside exposure, and the ability or lack of ability from the internal side, its resilience. In 1992, the International Panel of Climate Change, IPCC, defined vulnerability as the degree of incapability to cope with the consequences of climate change and sea-level rise, in 1996 IPCC, through Watson et al. (1996) defined it "as the extent to which climate change may damage or harm a system; it depends not only on a system's sensitivity but also on its ability to adapt to new climatic conditions".

Lewis (1999) definition of vulnerability is associated to shock and ability for resistance and recovery. Also in 1999, Klein and Nicholls express vulnerability for the natural environment as a function of three main components: resistance, resilience and susceptibility. Messner & Meyer (2006) and Merz et al. (2007), narrowed the definition of vulnerability to elements at risk, exposure (damage potential) and (loss) susceptibility, instead Mitchell (2002) expressed vulnerability as a function of exposure, resilience and resistance. A wide group of environmental researchers Kasperson et al, 2005; Adger 2006, Brooks, 2003; IPCC, 2007, see the vulnerability by combining the concepts above into a function of vulnerability related to exposure, sensitivity and resilience (adaptive capacity). Kelly and Adger (2000) social vulnerability to hazards is determined by their "existent state, that is, by their capacity" – or capacity to react and recover, and to deal with the everyday stresses. It is seen as the residual impacts of climate change after adaptation measures have been implemented (Downing, 2005). This definition includes the exposure, susceptibility, and the capability of a system to recover, to resist hazards as a result of climate change. Pelling 2003) definition of vulnerability is the exposure to risk and the incapacity to obviate potential harm.

Jones and Boer (2003), Sarewitz et al. (2003), Green (2004) have three (quite similar) definitions which are contemporaneous and express vulnerability as potential damage or harm caused to a system by an extreme event or hazard. Adger (2006) focuses on "shocks and stressors" near "capacity for adaptive action." Adger (2006) spotlighted the vulnerability as the state of susceptibility to harm from exposure to stresses associated with environmental and social change and from the absence of capacity to adapt.

McEntire, 2010, addresses the vulnerability through an integrated approach, McEntire takes into consideration two technocratic (Lewis and Mioch, 2005; Mileti, 1999) and two sociological schools (Bosher et al., 2007; Adger et al., 2003). The physical science school stresses living in safe areas, the engineering school concentrates on the built environment and ways to increase resistance through construction practices and methods of fabrication, the structural school concentrates on traditional notions of vulnerability more than the other three, and it stresses susceptibility based on socioeconomic factors and demographic characteristics and the organisational school stresses resilience or the effectiveness of response and recovery, while Hufschmidt, 2011, analyses the concepts of vulnerability through the view of "human ecologist school" and "structural view". Perrow (2006) specifies in his structural view of vulnerability, that the last one is determined by "economics and politics".

Cutter, (1993), Cutter (1996) defines vulnerability as a hazard which includes natural risks together with social response and action. Later on, Cutter (2005), Cutter et al. (2006) note, the difficulty to evaluate vulnerability, the links and mutual benefits among many systems to comprehend vulnerability to calamity. Burton & Cutter (2008) defined "vulnerability as the potential for loss and involves a combination of factors that determine the degree to which a person's life or livelihood is put at risk by a particular event". Downing, 1991, Wisner et al. 2004 and McEntire, 2008 define vulnerability as an evaluation of exposure to harmful occurrence and probability that individuals will lack the capacity to rebound.

In the past United Nations (1979) have defined flood vulnerability as the degree of loss to a given element, or a set of such elements, at risk resulting from a flood of given magnitude and expressed on a scale from 0 (no damage) to 1 (total damage). Since the quantification of vulnerability can help in decision making processes, parameters and indicators (indices) should be designed to produce information for specific target areas and they should provide information to counter different hazards which societies face, like floods (Douben, 2006; Page, 2000; Vaz, 2000; Mirza, 2003; Davidson, 2004).

In 2005, Veen & Logtmeijer broaden the concept of vulnerability to explain flood vulnerability from an economic point of view. Vulnerability to floods is defined as the extent to which a system is susceptible (McCarthy et al., 2001) to floods due to exposure, a perturbation, in conjunction with its ability (or inability) to cope, recover, or basically adapt (Balica et al, 2009). Gheorghe (2005) explains vulnerability as a function of susceptibility, resilience, and state of knowledge.

In socio-economic science Ramade (1989) includes in his approach of vulnerability, human and socio-economic terms; involving the predisposition of goods, people, buildings, infrastructures and activities to be damaged, offering low resistance, as it was introduced in the 1980s in some geographical studies. These latter studies interpreted the vulnerability of a geographical or territorial system as the result of different behaviour and coping capacities in socially, economically and technologically heterogeneous contexts (Menoni, 1997). Watts and Bohle (1993) analyse social vulnerability in the context of hazards and responses of communities to deal with resistance and resilience. The social vulnerability is intrinsically tied to several related processes the fragility, the susceptibility and lack of resilience of the exposed elements (Cardona, 2003). The author calls the exposure, physical fragility and tries to holistically integrate the contributions of physical and social sciences to define a vision of indicators which create vulnerability. Vulnerability is the degree of fragility of a (natural or socio-economic) community or a (natural or socio-economic) system toward natural hazards (EPSON, 2006).

Perry (2006), Quarantelli (2005) and Sorenson and Sorensen (2006) highlight how vulnerability is a socially constructed issue, the social inequalities divide people in vulnerable or not.

As a dissimilarity to simulation-models, the indicator based approach is uncomplicated (frequently linear) and not explicit in time, but can be a predictive tool.

2.2.1. Vulnerability Indicators

An indicator, or set of indicators, can be defined as an inherent characteristic that quantitatively estimates the condition of a system. Gomez (2001) states that 'they should be focused on small, quantifiable, understandable, unambiguous and telling pieces of a system that can give people a sense of the bigger picture', and these requirements have been reiterated by others (De Bruijn, 2004; Merz et al., 2007).

Indicators are simple numbers which express reality, for example: the rate of mortality in a region, the GDP per capita, the storage capacity of a dam. Understanding each concept and considering certain indicators may help to characterize the vulnerability of different systems, by which certain actions can be identified to decrease it. Some indicators, such as the population readily available for deployment in flood risk management, as well as the suitability of flood protection measures and flood risk management organisations or institutions, can only be measured during flood events (Messner and Meyer, 2006; Penning-Rowsell and Wilson, 2006).

The first step in any indicator-based vulnerability assessment is to select indicators (Sullivan, 2002; Sullivan and Meigh, 2005) while keeping their number to a minimum. For example, in a study on climate change (Canadian Council of Ministers of the Environment, 2003), nearly 100 indicators were considered. However, only 12 indicators were retained (Climate Vulnerability Index, (CVI)). These were grouped into two sections: those impacting more directly on nature and those impacting more directly on people (IPCC, 2001).

Two general approaches are used for indicator choice; the first one is based on a theoretical comprehension and the second one is based on statistical relationships. Theoretical comprehension has a function in both approaches. The first one corresponds to a deductive research approach and the second one to inductive research approach.

The standard practice, using a deductive approach, is to follow a conceptual framework (i.e. understanding the phenomena, identifying processes and then selecting indicators) or to assemble a list of indicators using criteria such as suitability, usefulness and ease of recollection. The deductive approach of choosing indicators engages relationships derived from theoretical structure then choosing indicators on their basis. In the deductive research approach, verification engages assessment of the goodness of fit between theoretical forecasts and experimental facts (Adger et al., 2004).

Recognizing the deficiencies and the inherent capacity for a better indicator selection and such evaluations is decisive. From the results can be seen the limitations or the enhancement of the indicator selection steps, as well as notions of vulnerability, conceptual approaches, weighting, conceptualization, data gathering and analysis.

In working with an indicator-based methodology, it is important to bear in mind two aspects: firstly, that local capacity and vulnerability, are shaped by processes, whether social, economic, environmental and physical, and thus vary dynamically both in time and in space; and secondly, that population, homes and communities may be faced by various strains (related to flood vulnerability) at the same time (De Waal, 1989), such as population growth, economic change, political conflict or climate change.

Inductive research frequently uses experimental overviews, along with observed content and statement of experimental regularities. The premise consists of overviews resulting by induction from data, the determinations of practices in data that can be generalized.

Indicators provide information about the system´s elevation, location, population density, land-use, their closeness to the stream, their closeness to inundation areas or return periods (frequency of occurrence) of different types of floods in the floodplain, as well as the awareness and preparedness of the social and physical system.

Awareness and preparedness indicators for population and their surroundings show the awareness of people for coping with hazardous events. For instance the number of houses protected by structural measures, like dikes or dams, the number of houses with flood insurance, etc. These measures can only be taken before flood events occur. Other indicators, such us the emergency services in disaster management, as well as the quality of flood structural and non-structural measures and disaster management organisations or institutions, can be measured only during flood events, (Messner & Meyer, 2005).

The capacity of individuals and communities to handle the impact of floods is often linked to socio-economic indicators. These indicators are in general information on age, gender, race, poverty, social relations, education, disable people, children less than 16, elderly more than 65 and institutional development (e.g., Smith, 2001, Blaikie et al. 1994, Sultana, 2010, Watts/Bohle, 1993).

2.2.2. Vulnerability Indices - How composite vulnerability indices are constructed

This chapter besides exposing different vulnerability's characteristics advocates composite vulnerability indices based on indicators.

In 1920 the use of indices as policy tools started (Edgeworth, 1925; Fisher, 1922). Indices are numbers based on indicators which assess a quantity relative to a base period, (Sullivan, 2002). An indicator, or set of indicators, can be defined as an inherent characteristic which quantitatively estimates the condition of a system; they usually focus on minor, feasible, palpable and telling piece of a system that can offer individuals a sense of the bigger representation. The indicators play a gradually more significant policy role; also they represent only succinct sides of a system at the diverse spatial scales. The first step in an indicator-based vulnerability assessment is to select indicators. The benchmark is to gather a list of proxies using the following criteria: suitability, definitions or the theoretical structure, availability of data.

The two standard procedures to select indicators are; the first one is based on a theoretical comprehension of relationships and one is based on statistical understanding. Theoretical comprehension has a role in both. One approach is characterized by the deductive research approach and the other is based on inductive research approach.

The selecting indicators approach, the deductive one, involves recommending links resultant from theoretical framework and choosing proxies based on these links. In the deductive research approach, verification involves assessment of the goodness of fit between theoretical predictions and empirical evidence (Adger et al., 2004).

Inductive research frequently uses observed generalizations, filled with pragmatic content and statement of empirical regularities. Hypothesis consists of generalizations resulting by induction from data, the result of examples in data that can be generalized.

Scholars concerned with the structural factors behind vulnerability for specific spatial scales should check the scores and rankings of these in the individual indicators. Those which need to spot very vulnerable systems for reasons of adaptation support will consider composite vulnerability indices practical.

Restraint should be considered while selecting the key vulnerability indicators, or while understanding them to create composite vulnerability indices. As Kaufmann et al., 1999a and Kaufmann et al., 1999b, mentioned, "is a high degree of heterogeneity in the way the data/indicators have to be collected". This admonishing point must be considered when integrating the data in a composite indicator based methodology.

Description of vulnerability indices – different approaches of vulnerability indices based on natural hazards

Economic Vulnerability Indices
In 1985, a conference in Malta brought Lino Briguglio the idea of constructing the first Vulnerability Index (VI) to assess economic vulnerabilities for small countries. In 1992 the index was in fact developed in the wrap-up of the Barbados Global Conference on the Sustainable Development of Small Island Developing States, as a tool to draw the attention of the international community to the vulnerability of SIDS.

Several versions were created for this index, mainly by Briguglio (1992, 1993, 1995, 1997, 2003, 2004), the Commonwealth Secretariat and Crowards (1998 and 1999), Atkins et al (1998 and 2001), a "commonwealth" vulnerability index for developing countries: the position of small states, Easter (1998) Chander (1996) and Wells (1997). By far all Economic Vulnerability Indices reach the conclusion that small states (SIDS) are amongst the most vulnerable countries.

The Composite Vulnerability Index for Small Island States (CVISIS, Briguglio, 2003, 2004)
The aim of the index is to point out the intrinsic vulnerability of such states in comparison to large countries which possess several advantages associated with their large scale.

This index was composed of four indicators: a two-level indicator, a small or large state, with numerical values 1 or 0 respectively; the vulnerability or susceptibility of the country in relation to natural disasters; the economic exposure of the country, "export dependence, the average exports of goods and non-factor services as a percentage of the GDP"; and the need of variegation, the UNCTAD diversification index.

Through the use of weighted least squares routines, the index was represented mathematically through the following equation:

$$\text{CVISIS} = 1.4142 + 0.0096 \text{ Vul x D} + 0.0322 \text{ Ex-Dep} + 3.3442 \text{ Div} \qquad 2.1$$

In this equation:
> Vul represents the susceptibility of the country to natural disasters;
> D is a two level indicator for the respective country regarding its status as a small state;
> Ex-Dep represents the economic exposure of the country;
> Div stands for the lack of diversification in a particular country.

The selection of weights was carried out using regression techniques and eliminating extreme values that might shift the index in undesired directions. Of the 111 countries (both small and large) over which the index was assessed, 11 were eliminated on this issue of extremes values (Villagran, 2006).

An updating and augmenting EcVI was done by Briguglio and Galea (2011). The four above indicators are becoming "(a) economic openness, (b) export concentration (c) peripherality and (d) dependence on imports". To compute the EcVI these components are adding up. In order to standardise the index the following formula was used (Briguglio and Cowards): (Xi – Min X) / (Max X – Min X), the range of results is between 0 and 1 (the most vulnerable).

As well in 2011, the French school through Guillaumont came up with a wide and wise hierarchical description of the EcVI development. The EVI (economic) of Guillaumont, (2004a) presents some improvements, here the EVI (economic) is an aggregating index computed on seven component indices, four shock and three exposure indices. With a mathematic averaging, the same weight is given shock indices and to the exposure indices. In the shock indices, the same procedure is used to natural and outside shocks, meanwhile in the exposure indices equal weight is given to population size and to the other indices.

$$\text{EVI (economic)} = \text{sqr}[1 - (1 - \text{EXP})(1 - \text{SK})] \qquad 2.2$$

Guillaumont, 2011 presents the results of the comparison between Small Island Developing States and Least Developed Countries, showing that EVI (economic) "on average is not only higher in the LDCs than in any other group of countries (except SIDS), but also does not appear to have declined, as in other groups (SIDS included)".

Global Risk and Vulnerability Index (UNEP, 2004)

The United Nations Environment Programme (UNEP) Division of Early Warning and Assessment (DEWA) and GRID-Geneva are developing a Disasters Risk Index under their Global Risk and Vulnerability Trends per Year (GRAVITY) project. An index which will show inter-country comparisons, and it is built on GRID-Geneva's Project for Risk Evaluation, Vulnerability, Information and Early Warning (PREVIEW).

The GRAVITY project examines the major hazard types: droughts, floods, cyclones, volcanoes, earthquakes and windstorms.

A main aim of the PREVIEW project is to create indices of human exposure to all hazards, using GIS data. The theoretical framework considered by UNEP is given by:

$$\text{Risk} = \text{frequency} \times \text{population} \times \text{vulnerability} \qquad \qquad 2.3$$

Where:

"Risk = number of expected human losses / exposed population / time period;
Frequency = expected (or average) number of events / time period;
Population = number of people exposed to hazard;
Vulnerability = expected percentage of population loss due to socio-political-economic context".

Global Risk and Vulnerability Index Trends per Year (GRAVITY), describe the concepts, data and methods applied to achieve the Disaster Risk Index (DRI). Categories of potential vulnerability indicators were defined as (Peduzzi et al., 2001): economy; dependency and quality of the environment; demography; health and sanitation; politics; infrastructure; early warning and capacity of response; education; development.

Vulnerability indicator data used in GRAVITY are (Peduzzi et al., 2001):

> ➢ An urbanization indicator was selected in order to include the fact that urban populations may be more or less exposed to a hazard than other populations, depending on the hazard. Urbanization is considered an indicator of affected population.
> ➢ An indicator of corruption was included in the selection, for it might contain information about presence of dangerous situations, e.g. houses built in hazardous areas. Hence, corruption is an indicator of vulnerability.
> ➢ The Human Development Index was selected because it seems rather natural to assume that there is a strong correlation between a country's development level and its mitigation capacities. Note that nor life expectancy or literacy rate were selected in the set of vulnerability factors. The reason is that life expectancy and literacy rate were strongly correlated, and that HDI provides even more information by itself.
> ➢ Population density is an indicator of affected population. Exposure is important for a given hazard if population is concentrated.
> ➢ It is assumed that GDP/capita is an indicator of mitigation capacity.
> ➢ Urban growth over last 3 years. The assumption is made that fast urban growth may result in poor quality housing, thus making people more vulnerable. However, this assumption may very well be only valid in particular regions. Yearly urban growth was not used because of its high variability. Considering growth over a longer time span is

certainly more likely to represent a risky housing situation. In that context, urbang3 is considered as an indicator of vulnerability.

➢ Population growth over last 3 years. The assumption is made that fast population growth may create pressure on housing capacities, and result in risky situations increasing vulnerability.

Considering a given disaster type: let Y be the vector of n observed damages, each element of vector Y corresponds to a different disaster that happened in a particular country c at a particular time t

$$Y = [victimsict]i=1,...,n$$

and let X the matrix of vulnerability factors corresponding to the country and time (when possible) of yict, X=[x1i ; x2i ; ... ; x7i]i=1,...,n

Where:

x1=popdct, x2=corupc2000, x3=hdic1998, x4=gdpcapct, x5=urbanct, x6=urbang3ct, x7=popg3ct

The following linear regression model is proposed $Y=\beta \cdot X + \varepsilon$

Where β is the vector of parameters β'=[β1 ; β2; ... ; β7]

and ε is a random perturbation satisfying the usual hypothesis of classical linear regression models.

Disaster Risk Index (UNDP, 2004)

The UNDP, 2004, developed a Disaster Risk Index, DRI, which quantifies and compares the degrees of physical exposure to vulnerability, hazard and risk on a country by country with respect to four hazards: floods, earthquakes, cyclones and droughts. The DRI is a mortality calibrated index. The DRI indexes the countries based on the hazard, the level of physical exposure, the level of relative vulnerability and the risk level. In the DRI, vulnerability cites the indicators less able to absorb the impact and recover from a hazard event at population level. These may be social, economic, environmental and technical. This index uses 26 indicators.

In 1979, UNDRO defined risk as the frequency of the hazard multiplied with the elements at risk, population exposed and vulnerability. In DRI the hazard's frequency of hazard and exposure population is named physical exposure, so the risk is a combination of physical exposure and vulnerability. The vulnerability in DRI is defined as the average number of deaths per exposed population. The DRI offers for each type of hazard a different formula using different indicators. For example for floods (Dao & Peduzzi, 2004):

$$\ln(K) = 0.78\ln(PhExp) - 0.45\ln(GDP_{Cap}) - 0.15\ln(D) - 5.22 \qquad 2.4$$

Where: K = people killed in the flood event
PhExp = average number of people exposed to flood event
GDP/cap = the normalised Gross Domestic Product / capita (purchasing power parity)
D = population density (i.e. the inhabitants affected divided by the affected area)

The DRI should be seen as a group of indicators that show the countries which are most at risk, vulnerable and exposed to floods, earthquakes, cyclones and droughts.

Environmental Vulnerability Index (Pratt et al., 2004)

The Environmental Vulnerability Index (EnVI), Pratt et al., 2004, was developed by the South Pacific Applied Geosciences Commission (SOPAC); the aim of the EnVI is as the EcVI to show the vulnerability of small island developing states (SIDS) from a series of natural and anthropogenic hazards, the index is based on 50 indicators; these indicators represent environmental integrity or degradation, risk and resilience.

The EnVI is naming its indicators as "smart indicators"; the authors (Pratt et al., 2004) use 50 indicators "which aim to capture a large number of elements in a complex interactive system while simultaneously showing how the value obtained relates to some ideal condition" (UNEP, 2004). TheEnVI indicators focus on the preeminent scientific comprehension presently available and have been developed after discussion sessions with country experts, international experts, interest groups and other agencies. The indicators are classified into 5 classes (Kaly et al., 1999): M = Meteorological; G = Geological; B = Biological; C = Country Characteristics; and A = Anthropogenic.

The 50 indicators selected to measure environmental vulnerability are classified into a range of sub-indices including: hazards, resistance, damage, climate change, biodiversity, water, agriculture and fisheries, human health aspects, desertification, and exposure to natural disasters. These indicators can be grouped into three sub-indices namely: REI = Exposure to human and natural risks per hazards; EDI= Environmental Degradation Index. The EnVI measures the present position of the 'health' of the environment. IRI = Intrinsic Resilience Index.

Environmental indicators are varied and include variables for which the responses are qualitative, numerical and on various scales (linear, non-linear, or with different ranges). Several different indicators are used resulting in a wide variety of different unit measurements.

The indicators are chosen based on expert judgment; they are heterogeneous and their resulting values are rated on a scale of 1 to 7, with 7 representing high vulnerability, an overall average of all is calculated to generate a country's EnVI. The index has been applied to a limited number of SIDS to date.

Coastal Vulnerability Index (Gornitz and Kanciruk, 1989)

The first CVI was developed by Gornitz and Kanciruk (1989), Gornitz et al. (1991), Gornitz (1991) and then used by others Thieler and Hammar-Klose (1999), Shaw (1998), Doukakis, (2005), Abuodha & Woodroffe, (2007). In this index the six variables are related in a measurable way that manifests the relative vulnerability of the shore to physical changes due to sea-level rise. This technique emphasizes areas where the diverse effects of sea-level rise may be the peak, Shaw et al. (Sensitivity Index(SI)) (1998) & Carter (1990), SI for Ireland, Forbes et al. (2003), Erosion Hazard Index. The coastal vulnerability index (CVI) is computed as the square root of the multiplication of the ranked variables divided by the total number of variables; The vulnerability classification is based upon the relative contributions and interactions of six risk variables.

$$CVI = \sqrt{\frac{(a*b*c*d*e*f)}{6}}$$

2.5

where, a = geomorphology, b = shoreline erosion/accretion rate, c = coastal slope, d =relative sea-level rise rate, e = mean wave height, and f = mean tide range. The CVI ranges (low (0-0.25) – very high (0.75-

1.00)), each computed value falls into the relevant quartile and the coastal region is then characterized accordingly.

Coastal Vulnerability Index (McLaughlin and Cooper, 2010)

A multi-scale coastal vulnerability index developed by McLaughlin and Cooper (2010), uses a function of the "physical nature of the coast (which controls its ability to respond to perturbation), the nature (frequency and magnitude) of the perturbation (the forcing factor) and the degree to which such changes impact on human activities or property". The coastal vulnerability can then be shown by using three elements and 17 variables, 7 for coastal characteristics (solid and drift geology, shoreline type, river mouths, elevation, orientation, inland buffer), 4 for coastal forcing (significant wave height, tidal range, difference in storm and modal wave height, storm, frequency) and 6 for socio-economic factors (population, cultural heritage, roads, land use, railways and conservation status):

Vulnerability = function of coastal characteristics (resilience and susceptibility) + coastal forcing + socio-economic factors.

The total CVI is computed as a summed up of the three components and then divided by 3. The values are presented between a range of 1 to 5, as well as in the case of Gorintz, (1990), where 5 contributed most strongly to vulnerability and 1 contributed least.

Coastal Vulnerability Index (Pethick and Crooks, 2000)

In 2000, Pethick and Crooks, proposed "a simple and preliminary" coastal vulnerability index relating to relaxation time over the return interval. Eight different types of shoreline where used cliffs (Brunsden & Chandler (1996); Moon and Healy 1994), beaches (Bascom (1954); Gunton (1997)), sand dunes (Ritchie and Penland (1990); Orford et al. (1999)), mudflats (Pethick (1996)), spits (De Boer (1988)), salt marshes (Pethick (1992)), estuaries (Metcalfe et al. (2000)) and shingle ridges (Forbes et al. (1995); Orford et al. (1995)). A coastal vulnerability index built up from different coastal forms gives a first order suggestion of the sensitivity of the landform to slight changes in its environment.

Drought Vulnerability Index (Wilhelmi and Wilhite, 2002)

In 2002, Wilhelmi and Wilhite, assessed a vulnerability index to agricultural drought in Nebraska. The index focuses on four factors, two biophysical, soil and climate and two social, land use and irrigation. These factors were combined in ERDAS Imagine GIS in order to map vulnerability to droughts in Nebraska. A numerical weighting system was applied, to each factor a relative weight was given between 1 and 5, where 1 is least significant and 5 is the most significant. The result map held four classes of vulnerability: 'low', 'low-to- moderate', 'moderate', and 'high'. Another drought vulnerability index, (USAID/FEWS, 1994), a composite index of vulnerability was based on three factors: crop risk, market access, and coping strategies.

Climate Vulnerability Index (Sullivan and Meigh, 2003)

Climate Vulnerability Index (CVI), 2003, Sullivan and Meigh, climate change indicators help us to establish if our climate is changing or not. These indicators are stands on characteristics of climate, such us precipitation and temperature. Other indicators show whether or not a changing climate is distressing the environment and individuals lives.

The Climate Vulnerability Index is a holistic methodology of assessing water resources and maintains the sustainable livelihoods approach used by many donor organizations to assess development progress. The index ranges between 0 to 100; with the total being produced as a weighted mean of six key components. Every component is also scored from 0 to 100. The six major categories or components are Resource (R), Access (A), Capacity (C), Use (U), Environment (E) and Geospatial (G).

In order to assess the CVI in practice, geographical types were identified; each of these has particular aspects which make it vulnerable to climate variability and change.
The methodology used for CVI is based on the methodology of Water Poverty Index developed by Sullivan, 2002:

$$CVI = \frac{w_r R + w_a A + w_c C + w_u U + w_e E + w_g G}{w_r + w_a + w_c + w_u + w_e + w_g} \qquad , \qquad\qquad 2.6$$

Where: $w_r, w_a, w_u, w_c, w_e, w_g$ – the weights of indicators.

Every component is made up of sub-components; the components are joint using a composite index structure.

There are different vulnerabilities to climate change, some of those studied are vulnerability to climate related mortality, social vulnerability to climate change, some countries have even defined their vulnerability to climate change using several indicators; for example: Canada, Peru, USA, etc.

Mortality from climate-related disasters can be quantified via emergency actions database, statistical relations between mortality and select likely proxies for vulnerability are used to select key vulnerability indicators. Brooks et al (2005) selected 11 indicators: literacy rate; literacy rate, over 15 years; population with access to sanitation; maternal mortality; life expectancy at birth; 15-25 year olds; calorific intake; civil liberties & political rights; voice and accountability; government effectiveness literacy ratio (female or male).

The indicators can be separated in three categories: Governance; Health status; Education.

Almost 100 possible indicators were examined for climate change report in Canada (Canada Council of Ministers of the Environment, 2003). The 12 indicators which remained were grouped into two sections (Nature: sea level rise, sea ice, river and lake ice, glaciers, polar bears, plant development and People: traditional way of life, drought, great lakes, frost and frost free season, heating and cooling, extreme weather). The first one includes those whose impacts are more directly on nature; the second, those whose impacts are more directly on people (IPCC, 2001).

Social vulnerability to climate change (Adger, 1999)
Adger (1999) describes another type of vulnerability; social vulnerability to climate change is the social exposure to stress as a result of societal and environmental changes. The author proposes a set of indicators to check the relative vulnerability of a sample of individuals or a social position. Among the indicators are:

- ➤ Poverty: income - an economic indicator of poverty;
- ➤ Resource dependency at the individual level;
- ➤ Inequality: as an indicator of collective social vulnerability, which affects directly the vulnerability through constraining the options of households and indirect through its links to poverty and others factors;
- ➤ Institutional adaptation at the collective level.

The CVI presents a influential method to analytically state the vulnerability of human communities in relation to water resources. The CVI approach integrates the physical, social, economic and environmental matters. The results are easy to understand –with a single number the vulnerability of a particular location can be express–meanwhile the essential data can be examined; these processes is transparent and open (Sullivan & Meigh, 2003). The CVI is appropriate for assessing vulnerability of climate variability, for assessing the impacts of climate change, all of this by combining climate scenarios with anticipated changes in social, economic, environmental and physical circumstances.

Climate Vulnerability Index (Yohe, 2006)
In 2006, Yohe et al. developed a set of indices of (aggregated outcome) vulnerability to climate change, after Brenkert and Malone, (2005) that is based on different assumptions regarding climate sensitivity, "the development of adaptive capacity", and other calibration parameters. The indices endure from fundamental methodological and conceptual limitations. The project website displays 144 global vulnerability maps.

The composite vulnerability indices of country i at time t is calculated as:

$$V_i(t) = \Delta T_i(t) / AC_i(t) \qquad\qquad 2.7$$

$\Delta T_i(t)$ is the predictable change in national average temperature (i.e., a rational-scaled variable) and $AC_i(t)$ is a normalized index of national adaptive capacity (i.e., an ordinal-scaled variable).

Social Vulnerability Index (Cutter et al., 2003)
The Social Vulnerability Index (SoVI), Cutter et al., developed in 2003, was used to characterize relative degrees of social vulnerability for counties in the United States, the dominant factors are: socio-economic status, development density, age, gender, race/ethnicity.

Initially, over 250 indicators were taken into account, but after testing for multi-collinearity among the indicators, 85 raw and calculated indicators were derived. Once the computations and normalization was done, only 42 autonomous indicators were considered in the statistical analyses. The method also helps duplication of the indicators at other spatial scales, thus making data assemblage more efficient. At the end 11 indicators were produced. The 11 indicators which remained are: personal wealth, age, density of the built environment, single-sector economic dependence, housing stock and tenancy, race (Asian and African-American), ethnicity (Hispanic and native American), occupation, and infrastructure dependence. This approach is based on the following algorithm: first standardisation of the indicators, perform PCA, then a selection of the components to be used (i.e. Kaiser Criterion), rotate initial solution (i.e. varimax and quartimax), interpretation and processing of the components, the combination of those components (i.e. equal weights, first component only) and lastly standardising index values.

Social Vulnerability to Climate Change for Africa (Adger and Vincent, 2005)
The indicators for Social Vulnerability to Climate Change for Africa (SVCC) Adger and Vincent, 2005, were chosen as a determinant of vulnerability. The indicators or surrogate indicators have been chosen within the limitations of data availability. The main indicators applied in the index are resulting from the World Bank which compiles about 800 World Development Indicators from data derived, either directly or indirectly, from official statistical systems organized and financed by national governments.

The process of developing indicators involves uncertainty at several levels. Adger & Vincent (2005), Vincent, (2004), present a social vulnerability index (SVI) to illustrate the issues of uncertainty in adaptive capacity.

The SVI is an aggregate index of human vulnerability to climate change-induced changes in water availability that is based on the weighted average of five composite sub-indices, Economic wellbeing and stability (20%), demographic structure (20%), global inter-connectivity (10%), institutional stability and wellbeing (40%) and nature resource dependence (10%).

The SVI is calculated through a simple equation (Villagran, 2006):

$$SVI = 0.2 \, Iewb + 0.2 \, Ids + 0.4 \, Iis + 0.1 \, Igi + 0.1 \, Inrd \qquad\qquad 2.8$$

In this equation:

> Iewb is the indicator associated to economic well being;
> Ids is the indicator related to demographic structure;
> Iis is the indicator associated to institutional stability;
> Igi is the indicator related to global interconnectivity;
> Inrd is the indicator associated to natural resource dependence.

The weights have been assigned to each indicator via suggestions emanating from an expert group. Most of the data has been acquired from international sources such as the World Bank, UN agencies, ITU, and Transparency International.

Vulnerability assessments to aquifers

Since in many areas of the world groundwater is one of the most important drinking water resource and its defense and sustainable use is of primary significance. Therefore, Neukum et al., 2007, describe four methods which assess karst vulnerability of the topmost aquifer. GLA (Hölting et al. 1995), method assesses the vulnerability by evaluating "the transit time of the percolation water" via the following factors: "the thickness of each layer in the unsaturated zone, the permeability of each stratum of the unsaturated zone and the amount of percolation water". DRASTIC (Aller et al. 1987), considers seven factors to assess vulnerability "depth to water table, net recharge, aquifer media, soil media, topography, impact of the unsaturated zone media and hydraulic conductivity of the aquifer". As other methods this one also considers weights (1 to the most important 5), ranges (according to the significance for groundwater vulnerability) and ratings (1 to 10, the most vulnerable). The ratings differ among 1 and 10 with 10 the most vulnerable case. The EPIK (Doerfliger and Zwahlen 1998), method was created to find groundwater protection zones in karst areas. This method considers four factors: "development of the epikarst, effectiveness of the protective cover, infiltration conditions and development of the karst network". The weighing system is between 1 to 3, the most important. The index shows that the upper values stand for higher protection.

The PI (Goldscheider et al. 2000) method estimates the "thickness and permeability of each stratum and an assessment of the degree of bypassing of the protective cover by surface and near surface flow which occurs especially within the catchment of sinking streams". These two factors are multiplied and will expose the spatial distribution of the protective purpose.

Water Poverty Index (Sullivan, 2003)

Water Poverty Index (WPI), developed by Sullivan, 2003, is a "holistic tool to measure water stress at the household and community levels". The index is based on five components: "access (population with access to safe water, with access to sanitation, fraction of land irrigated), resources (groundwater, surface water, and annual average precipitation), capacity (GDP, mortality rate, education Index), environment (indices of water quality, water stress, informational capacity, regulation and management capacity and biodiversity) and use (domestic, industrial, agricultural)". The index is applicable at country level. All the indicators are standardised using Eq.

$$I = \frac{X_i - X_{min}}{X_{max} - X_{min}}$$

2.9

The following equation determines the WPI:

$$WPI = \sum_{i=1}^{N} w_i * X_i$$

2 .10

Where:

I = Resource, Access, Capacity and Use

X i= Real value for country i

X max = The highest value country

X min = The lowest value country

WPI = Water Poverty Index

w i = Component's weight (20)

The index shows a country's relative position and lies between 0 and 100. The index approach was developed during pilot projects in Tanzania, South Africa and Sri Lanka, but was applied to 147 countries.

2.3. Perception of Flood Vulnerability

In the above mentioned vulnerability definitions, the hazards differ from definition to definition. Some of them give a definition of vulnerability to certain hazards like climate change (IPCC, 1992, 1996 and 2001) or environmental hazards (Blaikie et al., 1994); (Klein and Nicholls, 1999), (ISDR, 2004), but more important for this research is the definition of flood vulnerability.

In the past United Nations (1982) have defined flood vulnerability as the degree of loss to a given element, or a set of such elements, at risk resulting from a flood of given magnitude and expressed on a scale from 0 (no damage) to 1 (total damage). This definition falls short on this thesis focus, since it only considers some aspects of importance in the study of flood vulnerability.

Since the quantification of vulnerability can help in decision making processes, parameters and indicators (indices) should be designed to produce information for specific target areas and they should provide information to counter different hazards which societies face, like floods. In recent years the impacts of floods have gained importance because of the large number of people, economic activities and ecosystems that are impacted by their adverse effects.

Societies have developed close to water access, forcing its people to search for innovative ways to control and prosper with the more limited resources as the population grows, adding pressure on the water resources. The results of these solutions have been an important distinction on the development of societies, creating different problems for developed and developing countries.

Societies in the developed countries are well organized, their innovations harms the river system; most of them are heavily engineered, confined and leveed, safety standards are basically sufficient to prevent floods (Douben, 2006a). Society's vulnerability to floods is mainly reflected by possible economic losses as development grows; the cities develop in flood prone areas, leading to large economic growth and, thus increasing their vulnerability to floods.

The damages will be extremely high when a flood defence structure fails, especially in urbanized areas, where the most important industries are located. For example; an interruption of electricity caused by flooding will disrupt the system from its normality, and the economic damages will be enormous. In developed countries the losses will be reflected mostly in the economy, unfortunately few lives are spared.

Developing countries are characterized by widespread poverty, high population density, high rates of unemployment, pressure on rural land, illiteracy, and an economy usually dominated by agriculture and dependant on developed countries.

The developing countries vulnerability to floods can be reflected by these factors:
1. Socio-economic circumstances, high poverty level and lack of development;
2. Most of the infrastructure, including dams is not multipurpose (Page, 2000);
3. During floods, the use of inadequate measures are taken, such us: planning, design and their implementation (Vaz, 2000);
4. Countryside regions are heavily depending on agriculture and this affects more the economy, than urban areas;
5. Lack of education and prevention;
6. Lack of non-structural measures;
7. It is also a deficiency of adequate human and material resources to tackle the enormous floods disaster that happened in the past (Mirza, 2003).

Because of their vulnerability often millions of people become homeless and hundreds of thousands are in need of food and medicines, especially important is the high number of infected persons during floods, in developing countries. Houses, industries, infrastructure and agriculture are highly vulnerable. In these societies the losses due to floods are mainly lives, cultural damage, agricultural fields and cattle; the reconstruction costs are vast, and they usually take along time to recover, depending mainly on international aid (Davidson, 2004). All societies are vulnerable to floods, under different cases and situations, which make them somewhat unique; understanding the distinctions amongst them, may help to plan ahead and provide policy ideas to improve the quality of life of the people living in them.

A practice in defining vulnerability comes from natural hazards, such as floods: the extent to which a system is susceptible to floods due to exposure, a perturbation, in conjunction with its ability (or inability) to cope, recover, or basically adapt.

2.3.1. Flood Vulnerability Factors

Water resource systems are vulnerable to floods due to three main factors; exposure, susceptibility and resilience. The system's vulnerability (considering all spatial scales) involved the exposure and susceptibility of that system to hazardous conditions and the capacity or resilience/resistance of the system to deal with, adapt and/or recover from the effects of those conditions (Smit & Wandel, 2006).

2.3.1.1. Exposure

The values which are present at the location where floods can occur are the values which are exposed; values such us: cultural heritage, infrastructure, goods, agricultural fields or mostly people. Exposure is the extent to which humans and their homes are positioned in flood risk areas, (UNDP/BCPR, 2004). Exposure is generally described as patterns and processes which estimate its intensity and duration.

The indicators for this component can be separated in two categories; the first one covers the exposure of different elements at risk and the second one gives details of the general characteristics of the flood. The first category of indicators supplies information about the location, elevation, population density, land-use, their proximity to the river and to flooded areas. The second category provides information about return periods (frequency of occurrence) of different types of floods in the floodplain and similar to. All the indicators above inform us about the frequency of occurrence of inundations in floodplains, their duration and magnitude.

Exposure indicators provide specific facts about hazardous threats to the diverse elements at risk (Messner & Meyer, 2005).

In this thesis, exposure is defined as the predisposition of a system to be disrupted by a flooding event due to its location in the same area of influence (Balica, 2007).

2.3.1.2. Susceptibility

The concept of susceptibility, or sensitivity, has developed through the years. An important definition, Penning-Rowsell and Chatterton, 1977, focuses on the relative damageability of property and materials during floods or other hazardous events. The IPCC (2001) argued susceptibility as the affected system's degree, by climate related stimuli. At this moment the definition is still argued and creates confusion between social and natural scientists (Gallopin, 2006).

For Di Mauro (2006), susceptibility integrates the probability of a hazardous event, the differential exposure and the possible sensitivity of an objective. i.e. the extent to which a system could be potentially harmed or affected by a given hazard and the already existing ability of this target that could potentially diminish the level of damage.

Susceptibility relates to system characteristics, including the societal condition of flood damage development. Particularly the preparedness and awareness of affected population concerning the risk they live with (previous to flood), the institutions that are involved in mitigating and reducing the effects of the hazards and the existence of possible measures, like evacuation routes to be used during the floods.

For this thesis susceptibility is understood as the elements exposed within the system, which influence the probabilities of being harmed at times of hazardous floods.

2.3.1.3. Resilience

The concept of resilience and the related concept of resistance, used in ecology, are used to describe a system's ability to deal with perturbations and to continue without huge irreversible changes in their most important characteristics. Resistance is defined as the ability of this system to prevent floods, while de Bruijn (2005) defines resilience "as the ability of the system to recover from floods".

During the 1990s, the results of studies on complex systems influenced the concept of vulnerability, stressing the relation between vulnerability and resilience of a system and providing new theoretical tools for vulnerability studies (Galderisi et al. 2005).

Originally, the concept of resilience was outlined by Holling in 1973 as "a measure of persistence of systems and their ability to absorb change and disturbance and still maintain the same relationships between populations and state variables", a definition in tune with social science, but still significant for this thesis. Another definition (the second) of resilience is the capacity of any system to re-gain its equilibrium after a reaction to a disturbance (Begon et al., 1996; Jørgensen, 1992; Pérez España & Arreguín Sánchez, 1999). Holling states that the most essential feature of ecosystems is that they recover from disturbances. This recovery means that the principal characteristics of the system are restored, not that the exact same situation returns. Holling introduces the concept of resilience in addition to existing concepts within the systems approach in order to emphasize that systems are not stable and do not return to a stable equilibrium. This stability is, however, assumed in the second definition above. Therefore, Holling (1973) used this second definition of resilience for stability and not for resilience. The users of the second definition, however, consider both resilience and resistance as characteristics that make a system stable. A human being could, for example, survive floods in flood prone areas by having resistant waterproof house foundation (exhibiting resistance), or alternatively it could be flooded and recover while and/or after flood happens (exhibiting resilience). In stable environments, places where floods happen with a rare frequency, but big magnitude, more resistant types of houses (ability to endure changes due to that flood) will be found, whereas in very dynamic environments, such as coasts and natural floodplains, resilient houses dominate.

Walker (2004) argued that resilience is "the capacity of a system to absorb disturbance and being reorganized while undergoing change, so as to still retain essentially the same function, structure, identity and feedbacks".

Resilience is the capacity of all systems, i.e. a society or community, potentially exposed to hazards to adapt to any change, by resisting or modifying itself, in order to maintain or to achieve an acceptable level of functioning and structure (Galderisi et al, 2006). Pelling (2003) defines resilience also as the capacity to adapt, to adjust to threats and mitigate or avoid harm.

Resilience to flood damages can be considered only in places with past events, since the main focus is on the experiences encountered during and after the floods. Floods are a physical disruption which threatens social, economic and/or environmental systems. Flood resilience can be expressed as the ability of a system or community to defy or alter itself so that the harm of floods is mitigated or minimized.

In this thesis resilience is defined as the capacity of a system to endure any perturbation, such as floods, maintaining significant levels of efficiency in its social, economical, environmental and physical components.

Resilience in the relationship with adaptability and vulnerability
In different disciplines, resilience is often confused with vulnerability and adaptability. To clear this up, the definitions used for these concepts in this thesis are explained and the relationships between the two concepts and resilience are discussed. The aim of this section is to explain that the concept of resilience can be a useful addition to the already existing concepts, since it is clearly different from the already existing concepts as it corresponds with a systems approach.

Resilience and adaptability
Resilience, as defined in this chapter, relates to the capacity to adapt, to adjust to threats and mitigate or avoid harm from short and long term disturbances (McFadden, 2001); and also relates to the system's persistence measure and their capacity to endure change and disturbance and with all of these to preserve the same relationships among individuals and state variables.

A system's ability to cope with or to adapt to sudden permanent changes is called adaptability. Therefore, in this thesis the adaptability is a part of resilience.

Resilience and vulnerability
Vulnerability, and the related concepts risk and hazard are widely used in flood risk management. A hazard is the trigger to a disaster while the concept of vulnerability determines whether or in what circumstances such a hazard will result in a disaster. The concept of vulnerability is used both as a quantitative and a qualitative concept. Vulnerability of flood system depends on the socio-economic, environmental and physical context of it. Different components within this vulnerability concept are the potential impacts which a flood may have and the recovery capacity to overcome those impacts.

Although the concepts of vulnerability and resilience have similarities, they can be also used differently. Resilience, together with resistance alone describes how a system reacts to a disturbance, such as flood, while vulnerability relates to why and how a (socio-economic, environmental and physical) system responds in a given way. The origin of the two concepts is also different. The resilience concept is derived from stability theories and theories of system dynamics, while the vulnerability concept is mainly used in social science.

The central issue in the both concepts of vulnerability and resilience is the reaction of people, their livelihoods and the whole system around them. Resilience is thus not a synonym for vulnerability. In this thesis, the use of the resilience concept as part of vulnerability is studied. The concept of vulnerability is important because it represents here not only social science's knowledge of why people's lives are affected by floods and of the importance of floods to people's life and also the effect of their adaptability/resilience to floods, but also the system itself is exposed, susceptible and resilient to floods. Resilience, it is implied an appropriate counterpart of vulnerability.

2.3.2. Flood Vulnerability Indices

During the last few decades, scientific evidence has pointed to a marked increase in frequency (CRED, 2008), intensity and economic effects of meteorological-related events such as floods. The objective to develop indices is to provide decision makers with tools to assess and analyse flood events.

Flood Vulnerability Index (Connor & Hiroki, 2005)

Connor & Hiroki, 2005, presented a methodology to calculate a Flood Vulnerability Index (FVI) for river basins, using eleven indicators divided in four components. The index uses two sub-indices for its computation; the human index, which corresponds to the social effects of floods; and the material which covers the economic effects of floods. The purpose of the FVI is to serve as a tool for assessing flood risk due to climate change in relation to underlying socio-economic conditions and management policies.

Out of 40 identified possible indicators only eleven were acknowledged by a group of over 50 participants during an event at the Asian Development Bank Water Week, 2004 (Manila), these eleven indicators are: frequency of heavy rainfall (I1) belonging to climate component (C); average slope (I2), urbanised area ration (I3) belonging to hydro-geological component (H); TV penetration rate (I4), literacy rate (I5), population rate under poverty (I6), years sustaining healthy life (I7), population in flooded area (I8), infant mortality rate (I9) belonging to socio-economic component (S) and investment amount for structural measures (I10), investment amount for non-structural measures (I11) belonging to countermeasures component (M).

The methodology was tested on river basins in Japan, where there is a lot of accessible information. Relatively easily available indicators were selected to facilitate the application of the method to other basins (114) in the world. Using the Japan data, the researchers used multi-linear regression analysis to calculate the weights of each indicator to the human and material FVI, based on number of casualties and material losses of past flood events the indicators reflected the actual vulnerability to floods of each river basin. The weights of the indicators were presented with the following equation:

$$FVI = C + H + S - M$$ 2.11

$$FVI = (3*I_1) + (3*I_2 + I_3) + (-I_4 - I_5 + I_6 - I_7 + I_8 + I_9) - (I_{10} + I_{11})$$ 2.12

The FVI values using this methodology oscillate between 0 and 1, where 1 means the highest flood vulnerability and 0 represent the lowest vulnerability to floods.

Flood Vulnerability Index applied for river basins in Philippines

The methodology was also tested in 18 river basins in the Philippines, where some indicators were added or changed because of lack of information.

The equation used for The Philippines:

$$FVI = \frac{w_c C + w_h H + w_s S}{w_m M}$$ 2.13

The methodology included a step of converting the indicators into non-dimensional units, by interpolating the maximum and minimum of the series of data obtained, using the Equation 11.

Using this methodology allows for comparison of a series of river basins, but comparisons between two different series, for example river basins from different countries, can be misleading since part of the comparison involves the interpolation of data, and not the value of the indicator itself.

Integrated Flood Vulnerability Index (Sebalh, 2010)
In 2010 Sebalh developed an Integrated Flood Vulnerability Index (IFVI), based on four components: social (total population, females, growing population, population density and unemployed), economic (housing stock, dwellings and other units, vehicles registration and industrial commercial), ecologic (biological reserve/protected areas and Natura 2000) and physical (flood extent 1999 and flood scenario 0.00m > 4.00m). The indicators were weighed, the most important ones received the higher weight. A matrix is used in order to rate the components of the IFVI method, the social component receives the maximum rate 56, while economic 33 and ecologic 11. The vulnerability was computed using the following equation:

$$Vul = \sum_{i=0}^{n} (v_i * w_i)$$

2.14

where:
Vul = Vulnerability of the component
vi = vulnerability factor i
n = number of indicators
wi = weight of vi value
Σwi = w (indicator)
W(indicator) = total weight of specific indicator

The total IFVI is an add up of three of the components: social, economic and ecologic.

Flood Vulnerability Index

An improved methodology to compute a *flood vulnerability* index (developed by Balica, 2007), based on indicators, was developed, aiming at assessing the conditions which favour flood damages at various levels: *river basin, sub-catchment and urban area*. This methodology can be used as a tool for decision making to direct investments in the most needed sectors. Its implementation could guide policy makers to analyse actions towards better coping with floods.

The methodology involves two concepts. First, vulnerability, which covers three related concepts called factors of vulnerability: *exposure, susceptibility* and *resilience*. The other concepts concern the actual flooding; understanding which elements of a system is suffering from this natural disaster, called components of vulnerability. Four main components of a system are recognized which are affected by flooding: *social, economic, environmental* and *physical components*. The interaction between the vulnerability factors and the vulnerability components serves as the base of the proposed methodology.

The developed methodology distinguishes different spatial scales of flood vulnerability: river basin, sub-catchment and urban area. This permits a detailed interpretation of specific indicators and pinpoints actions to diminish focal spots of flood vulnerability. The larger scales in international committees aim to identify and develop necessary action plans to deal with floods and flooding. The smaller scales aim to improve the (local) decision making process by selecting action plans to reduce vulnerability at local and regional levels.

The methodology has been applied at various spatial scales, which resulted in interesting observations on how vulnerability can be reflected by quantifiable indicators.

Alongside the FVI results were presented in standardized form for further comparison between components and the methodology (Connor & Hiroki, 2005) and also serve the purpose of easier interpretation. The formula used to standardize FVI values between 0 and 1 is presented as:

$$sFVI = \frac{FVI_{specific}}{\max_{i=1}^{n}(FVI)_i}$$

2.15

2.4. Flood risk expressions

Flood risk management in a constricted common sense is the procedure of managing the flood risk situation (Plate E., 2002), in a broader sense, includes the system's measures, which will diminish the flood risk. The characteristics of *flood risk management* include difficulty, different spatial scales, even trans-boundary river basins; inter temporal issues and conflicts of interests. (Kenyon W., 2007), made a short description of each method which assess flood risk "highlighting the specific strengths", examining the *citizen's juries*.

Expressions of risk, related to floods

Risk - "the probability that a hazard will turn into a disaster". Independently taken, the vulnerability and hazards do not define risk, if they join, they turn into a risk or, the probability that disaster will happen.

Risk = Vulnerability * Hazard   2.16

After the theoretical definition used by UNEP risk is expressed in equation (15). Regarding flooding, the notion of risk is one of the inner subjects. Risk can be also expressed "as the probability of occurrence of an event multiplied by then consequences of that event" (Bouma J.J. et al, 2005).

Risk = Probability*Effect   2.17

Barredo, et al (2007), used the following expression for evaluating flood risk:

Flood risk = f (Hazard, Exposure, Vulnerability), JRC   2.18

In the IPCC-TAR (2001), the expression of risk is as shown in equation (17); however, in the IPCC-AR4 (2007), risk is generally understood to be the product of the likelihood of an event and its consequences.

Risk = Adaptation – vulnerability (IPCC, 2001)   2.19

The risk (Plate E., 2002) is defined as:

$$RI(D) = \int_{0}^{\infty} K(x|D)f_x(x|D)dx$$

2.20

This risk expression (2.17) is computed as a consequence function $K(x|D)$ where x is the extent of the incident causing the loads (e.g. the flood water level); D is the vector of decisions, (e.g. the storage capacity of a dam or the height of a dike), that act upon the (usually unfavorable) consequences K (reducing the reference to D from here on) of any X incident.

In one of its articles (Hall, J.W., et al, 2003), speaks about a quantified river and coastal model for flood risk evaluation that takes clear description of the reliability of flood structural and non-structural measures and their modifying effect on flood risk. This model used England and Wales GIS databases.

2.5. Validation of Vulnerability Indices

Lately, many approaches to determine vulnerability at the large-scale have been presented. Still, little is known concerning the accuracy and validity of these vulnerability indices.

Measuring vulnerability is a requirement of the European Floods Directive 2007/60/EC, the flood risk management strategies should "focus on prevention, protection and preparedness", it is one of the aims to reduce risk from natural hazards. As the European Floods Directive, in Kyoto, 2005, the 'Hyogo Framework for Action 2005–2015: Building the resilience of Disaster Reduction (WCDR), requires the need to "identify, assess and monitor disaster risks' (UN 2005: 12). To achieve this goal, the Kyoto declaration stressed the development of an indicator based systems of disaster risk and vulnerability for multiple scales. Indicators, all through an index, can be a guide to understanding in a holistic way the current state of a system, also indicating the possible strategies to improve the functioning of the system. Vulnerability indicators are not something new; they have been used for different risk based assessment for different fields of study, like social, economic, environmental or engineering. Having an understanding of all these areas of study can complement even more the understanding of the correct functioning of a water system.

Indices are a statistical concept, presenting an indirect manner of measuring a given quantity or state, in fact a measure which allows for comparison over time. There is no general approach or model to quantify vulnerability. This chapter despite the state of the art of diverse vulnerability indices also briefly illustrates the way of evaluating vulnerability.

For the development of these indices, the research stressed the need to identify indicators which would represent in a better way reality. Apart of the EVI and GRAVITY, all the indices have different weights for each indicator used, evaluating this individual weight must be done in a way that the end result improves the perception of reality given by the index.

The indices exposed in this chapter are categorised according to several natural hazards identified. Indices exposed to all types of hazards, such us: EnVI, EcVI, DRI, GRAVITY, CVISIS, or exposed to floods, such us: IFVI, or coastal erosion and coastal floods, to droughts, to water poverty (WPI, EPIK, DRASTIC, PI, GLA), to climate change, CVI, SoCVI; etc. Throughout the diverse indices to natural hazards, different facets of vulnerability can be measured. In order to measure these facets, key issues should be taken into consideration to construct any index. As seen the vulnerability indices are based on a choice of components depending on the hazard, choice of indicators to fit into the components and choice of vulnerability concept from different scholars.

Despite these issues, the focus of an index however is to "measure the un-measurable" (Birkmann & Wisner, 2006), to quantify something which cannot be measured directly (e.g., vulnerability to floods, vulnerability to droughts) and to quantify changes (e.g., the Human Development Index, EcVI). The vulnerability indices given in this chapter fit the perception of an index which measures something indirectly, and which is build up of defined components.

Decision makers need such indices. Standardised vulnerability indices help in assessing and monitoring the "elements at risk" (Merz et al., 2007). This need involves a range of subjective decisions, the choice of indicators for example, but if the vulnerability concept is well defined and clarified the evaluation and interpretation of composite indices will bring us understanding of where to mitigate risk and where to focus investments. Still there is no standardised way to measure vulnerability (Bohle et al. 1994), the measurement depends of each hazard that occurs, its frequency and its intensity.

Several weaknesses of using indicators and creating indices should bring into discussion the validity of such an aggregated method. Indicators are a part of the real complex life reflection, which help comparisons on time and space of communities, societies, countries and river basins (including trans-boundary conditions).

The main weakness of an index is that a system of indicators can never represent an actual and complete image of the actual situation. Furthermore, when summarising a situation in any number of indicators, information is always lost. Therefore, it is of the upmost importance that the decision-makers take the actual local situation and trends into account when designing policies.
Other weaknesses in the evaluation vulnerability through indices include the difficulty of quantifying certain social and environmental indicators. Not all the significant processes are included. The indices do not allow for projections to the future, it only accounts for the current state of the study area. It also do not account for interactions between indicators. Finally, it could be difficult to collect the required information where data availability and collection is limited.

Vulnerability is always a call into question concept, being differently defined by scientists, but a concept which comprises a multitude of processes to mitigate risk. The many aspects of vulnerability are difficult to concentrate into an indicator. The indicators taken alone are considered subjective matters, aggregating indicators increase the subjectivity and would be even more difficult to evaluate and analyse. Aggregating indicators result into an index, the index ties scientists and decision-makers, for scientists the index needs to offer certainty in science, a validated index; from the policy-makers point of view is also greatly dependent of the choice of indicators at the lowest level, and there is a real risk that unapprised choices at this level sieve through and can direct to an invalid index. Indices to natural hazards always should be under continuous development, this in order to assure robust results, and therefore vulnerability indices utility.

The indicators can be validated by using the independent variables of the independent second data set and running a logistic regression model (Fekete, 2009). In contrast, the vulnerability indices are extremely difficult to validate. It is well disseminated the validation method to look at correlation with past disasters data (Pelling and Uitto, 2001; Easter 1999; Brooks and Adger; 2003; Crowards, 1999). Still, using historical occurrences of disasters and applying the index to "temporally-specific data might at least act as a means of validation for the structure of the index in explaining social vulnerability" (Vincent, 2004). Vulnerability changes in time, less vulnerable places years ago, may now be highly vulnerable to natural hazards. For example, the exposure is permanently increasing, i.e. sea level rise, heavy precipitations, landslides, global warming; natural resilience is not enough, due to the presumed economic crash the amount of investments will decrease, in special in developing countries. Therefore a validation to past disaster data in the case of a vulnerability index is not appropriate. Another point of view, realistic attempt to validate an index will be a comparison between other indices developed for the same type of natural disaster applied to the same scale, i.e. coastal vulnerability index Gorintz versus McLaughlin and Cooper vs. Balica et al. vs. Pethick and Crooks.

Evaluating vulnerability and validating vulnerability indices is of maximum importance since the state of vulnerability can increase the risk to natural hazards. As said, one of the requirements of the European Floods Directive 2007/60/EC is that flood risk management strategies should "focus on prevention, protection and preparedness". By 2013 EU member states "must develop flood hazard maps and flood risk maps for real risk of flooding". By 2015 flood risk management plans must be drawn up for these zones. For that reason, these plans are to include measures to reduce vulnerability to natural hazards and its potential consequences by focus particularly on reducing exposure, protection and preparedness. Therefore, vulnerability maps and better understanding risk perception is needed.

2.6. Uncertainty in flood vulnerability

Uncertainty is the outcome of vague knowledge e.g. where the probabilities and extent of each hazard and/or their related consequences are doubtful (de Bruijn, 2005).

Whether uncertainties are important in flood risk management depends on the effects they have on decisions. If the choices to be made are in doubt then the decision is uncertainty (Green, 2003b). The uncertainty originates from vague/absence of knowledge of the alternatives, of the consequences of the alternatives, of the later situation of the system, or of the decision criteria (Green, 2003b). To decrease uncertainties is only useful for decision making process, if it changes the ranking of alternative options. If the status of options is certain, regardless of lots of characteristics being uncertain, still a choice/decision can be made (Green, 2003b).

Variability in nature is one of the mainly significant uncertainties in flood risk management (de Bruijn, 2005). If vulnerability to flood will always be the similar, it would be better manageable.

To solve and avoid uncertainties in vulnerability always will be difficult. Still, choices/decisions have to be taken.

With high level of uncertainty, FVI cannot essentially reduce uncertainty, but could be an effective tool that would assist decision makers in evaluating the impacts of different scenarios. The FVI tool can assist in modelling different alternatives for action and thus enable decision makers in making educated decisions, which would increase the mitigation and adaptation to flood risks in the most efficient way.

The function of FVI in risk management and dealing with uncertainty is to support and help planners, politicians, decision makers or specific stakeholders such as builders of infrastructure, dams, roads, buildings etc to better recognize which measures that can be taken to decrease vulnerability before possible harm is realized. It helps them to prioritise which actions that are most important. It can also identify hotspots related to flood events where for instance specific measures might be taken. It can also function as a tool to spread information for awareness rising.

Although this would not necessarily reduce uncertainty could help to have a better grasp of possible scenarios and therefore ensure that adaptation measures are as relevant and targeted as possible. This would not particularly reduce either statistical, scenario or levels of ignorance but it could help create thresholds for action by determining the point at which flood vulnerability needs to be dealt with.

CHAPTER 3

Development of flood vulnerability indices at varying spatial scales, its implementation and dissemination

Parts of this chapter has been published as:

Balica, S.F., Douben, N., Wright, N.G., 2009, Flood Vulnerability Indices at varying spatial scales, Water Science and Technology – WST 60.10. pp 2571-2580

Balica S.F., Wright N.G., 2009, A network of knowledge on applying an indicator-based methodology for minimizing flood vulnerability, Hydrological Processes Journal 23, pp 2983-2986, Published online 25 August 2009 in Wiley InterScience (www.interscience.wiley.com). DOI: 10.1002/hyp.7424

3.1 Introduction

Firstly, this chapter discusses the development of the flood vulnerability index methodology (FVI) at different spatial scales, (river basin, sub-catchment and urban area) and secondly the applicability of the FVI methodology by using an automated calculation of the FVI implemented through a web management interface (PHP).

Although, the focus of this thesis is on the relationship between flood vulnerability index components, the factors of vulnerability and the spatial scales, a holistic understanding of the methodology is expressed by the identification of the correct selected indicators and their quantification and definition.

The development of the conceptual model is discussed in Sections 3.2 to 3.6. Section 3.2 discusses the water resource systems, and the three spatial scales used; Section 3.3 focuses on the development of the FVI approach, as a stepwise for three spatial scales river basin, sub-catchment and urban area; Section 3.4 explains the network of knowledge on how to apply an indicator-based methodology for minimising flood vulnerability.

Background
Human population worldwide is vulnerable to natural disasters. Such disasters are occurring with increased frequency as a consequence of socio-economic and land-use developments and due to increased climate variability. In recent years the impacts of floods have gained importance because of the increasing number of people who are exposed to its adverse effects.

In recent years, flood research and flood protection policy has been interacting not only with the technical aspects, but also with the social and socio-economic aspects, which gained in importance in recent decades due to expansive and intensified land use, raising damage potential in floodplain areas. Due to these actions, the flood protection towards flood risk management follows a trend where the focus is more on the non-structural measures in order to mitigate flood response. However, the scientific developments and improvements in the analysis of flood protection were mainly formed by civil engineers in the past, focusing on technical and financial aspects and neglecting the significance of socio-economic and environmental factors. Flood risk management will bring a new, more interdisciplinary and holistic view on flood management and policy, a flood vulnerability index (FVI) tool is necessary in order to make comparisons across the diverse spatial scales. For easy comparison purposes, vulnerability index is introduced, comprising of a set of indicators representing various aspects relevant to magnitude and range of impacts and damages of floods to communities and environment.

Therefore this chapter describes a methodology for using indicators to compute a FVI which is aimed at assessing the conditions which influence flood damage at various spatial scales: river basin, sub-catchment and urban area. The methodology developed distinguishes different characteristics at each identified spatial scale, thus allowing a more in-depth analysis and interpretation of local indicators. This also pinpoints local hotspots of flood vulnerability. The final results are presented by means of a standardised number, ranging from 0 to 1, which symbolises comparatively low or high flood vulnerability between the various spatial scales.

The FVI can be used by international river basin organisations to identify and develop action plans to deal with floods and flooding or on smaller scales to improve local decision-making processes by selecting measures to reduce vulnerability at local and regional levels.

The methodology has been applied to various case studies at different spatial scales. This leads to some interesting observations on how flood vulnerability can be reflected by quantifiable indicators across scales, e.g. the relationship between the flood vulnerability of a sub-catchment with its river basin or the weak relation between the flood vulnerability of an urban area with the sub-catchment or river basin which it belongs to.

Flood vulnerability assessment plays a key role in the area of risk management. Therefore, techniques that make this assessment more straightforward and at the same time improve the results are important. To easily manipulate the amount of data and the results of the FVI methodology, this chapter also presents an automated calculation of a flood vulnerability index implemented through a web management interface (PHP) that enhances the ability of decision makers to strategically guide investment and identify hotspots related to flood events in different regions of the world. To test the applicability of this methodology using this website, many case studies are required in order to cover the full range of cases in terms of scale such as river basin, sub-catchment, urban area and coastal city.

This requires prompt solutions with large amounts of data and this has led to the development of this automated tool to help organize, monitor, process and compare the data of different case studies. The aim is to create a network of knowledge between different institutions and universities in which this methodology is used. It is also hoped to encourage collaboration between the members of the network on managing flood vulnerability information and also promoting further studies on flood risk assessment at all scales.

3.2. Water resources systems

The water resources systems studied in this chapter are divided into interdependent sub-systems. The natural river sub-system, in which the physical and biological processes take place, the socio-economic sub-system, which includes the societal (human) activities related to the use of the natural river basin and the administrative and institutional sub-system, including legislation and regulation, where the decision-making, planning and management processes take place (van Beek, 2006).

Each of these three sub-systems is defined by certain conditions. The natural river sub-system is delimited by climate and (geo) physical conditions, the socio-economic sub-system is formed by demographic, social and economic conditions, and the administrative and institutional sub-system is formed and bounded by the constitutional, legal and political system.

3.2.1. The relationship between Floods and Water Resource

A systems approach, similar to the one used in ecology must be adopted in order to be able to apply resilience in the context of flood risk management and the system relevant for flood risk management must be defined.

Generally, river basins are affected by floods at three main scales, with boundaries depending on their spatial scale: the river basin, the sub-catchment and the urban area. A flood risk management system is defined geographically as the combination of these spatial scales and the four components of water resources system.

Floods distress four components: *social, economic, environmental and physical* of the water resources system (ISDR, 2004), each of them belongs to one of the sub-systems described before, and their interactions affect the possible short term and long term damages. The components can be assessed by different indicators to understand the vulnerability of the system to floods

The social and economic components comprise the socio-economic and the administrative and institutional sub-system, whereas the environmental and physical components are part of the natural river sub-system.

3.2.1.1. Components of vulnerability

The *social component*, the flooding affects the day to day lives of the population that belongs to the system. This component relates to two factors: on the one hand the presence of human beings which encompasses issues related to, for example, deficiencies in mobility of human beings associated with gender, age, or disabilities (van Beek, 2006); on the other hand floods can destroy houses, disrupt communication networks, or even kill people. Included in this component are the administrative arrangements of the society, consisting of institutions, organizations and authorities at their respective level.

The social component includes indicators which are measures and/or variables to describe the context, capacity, skills, knowledge, values, beliefs, and behaviours of individuals, households, organizations, and communities at various geographic scales. Social indicators are typically used to assess current conditions or achievements of social goals related to human health, housing, education levels, recreational opportunities, and social equity issues.

The *economic components* are related to income or issues which are inherent to economics that are predisposed to be affected (van Beek, 2005, Gallopin, 2006). Many economic activities which can be affected by flooding events, among them are adversely agriculture, fisheries, navigation, power production, industries, etc. The breakdown of these activities can influence the economic prosperity of a community, region or a country. In recent years floods have intensified due to e.g., lack of environmental awareness, creating even more damages to the ecosystems; if the flood water is polluted or if large sedimentation processes occur, ecological systems can be disrupted significantly (Haase, 2003).

The economical component illustrates the well-being of the region of study. These indicators must provide knowledge on the capacity to produce and distribute goods and services which may be vulnerable to floods. For example, developing countries are characterized by low income per capita, human resource deficiencies, lack of investment and finance and weak internal interlinkages. On the other hand, developed countries can be distinguished by large amounts of investment in mitigation and counter measures, high life expectancy, flood insurances, urban planning, etc. If economic development increases, potential flooding damages may also increase.

The *environmental component* continues to relate to the interrelation between the sector and the environment and the vulnerability associated with this interaction (Villagran, 2006). Activities such as deforestation, urbanization and industrialization have enhanced environmental degradation, creating effects like climate variability and sea level rise, increasing the potential occurrence of floods.

The environmental component includes indicators which refer to damages to the environment caused by flood events or manmade interferences which could increase the vulnerability of certain areas. Activities like industrialization, agriculture, urbanization, afforestation, deforestation, among others have been proven to create higher vulnerability to floods, which may also create even more environmental damages. Some of the indicators taken into consideration are groundwater level, land use for economic activities or for natural reserves, degraded area, percentage of urbanized area, forest change rate, etc.

The *physical component* comprises geo-morphological and climatic characteristics of the system, and different infrastructures, like channels, reservoirs, dams, weirs, levees which have shaped its physical conditions. The physical component relates to the predisposition of infrastructure to be damaged by a flooding event. The physical component tries to explain how the physical condition, either natural or manmade, can influence the vulnerability of a certain region to floods. Some indicators found are topography, heavy rainfall, evaporation rate, flood return periods, proximity to river, river discharge, flood water depth, flow velocity, sedimentation load, length of coast line, etc.

3.2.2. Different spatial scales

A direct and precise measurement of flood vulnerability is difficult, due to the lack of necessary data and because vulnerability is geographically and socially differentiated (Adger et al., 2004; Adger, 2006). An interesting aspect of vulnerability is that it can be examined at different levels and scales for different issues, for example, the relative position of a certain spatial and societal scale to flood vulnerability.

In this thesis, scale is taken to mean the unit of analysis that is located at different geographical positions (e.g. river basin and urban area) while level is taken to mean a different type of component (e.g. social and economic).

The spatial scales for vulnerability represent progressively smaller areas of focus. While river basin and sub-catchment are clearly hydrologically defined boundaries for assessment, urban crosses into a more socially defined category. Urban areas are densely populated, which may make them especially vulnerable to flood effects. However, the urban area is included here as a scale due to the need to analyse its degree of vulnerability to inform decision making that is often focused on a single urban entity. The difference in the nature of the definition of level and scale terms must be borne in mind when considering the results and their inter-comparison.

The understanding of flood vulnerability of different river basins starts with categorisation. The different categories which can be distinguished in a river basin are related to size and to inherent characteristics.

The advantages of identifying these categories can be summarised as:
- Vulnerability is geographically and socially differentiated. Any assessment at national level must take into account regional patterns of vulnerability within the country and the distribution of vulnerability within the national community (Adger et al., 2004);
- It is increasingly recognised that vulnerability is a dynamic characteristic, a function of the constant evolution of a complex of interactive processes (Leichenko and O'Brien, 2002);
- Spatial heterogeneity results in a more accurate description of reality;
- It includes differences in vulnerability components and vulnerability factors;
- Political and administrative division can either facilitate or impede the availability of data, according to certain scales. Data from river basins stretching out over more than one country will be more difficult to estimate; data from urban areas may vary from country data;
- The final results will be more applicable and understandable through accumulation of knowledge of how vulnerability is distributed and how it is developing throughout the world.

Dividing the FVI into spatial scales, into different components, and linking them with the factors of vulnerability, i.e. (see Chapter 2) can assist in identifying weak points of a flood defence system. Hence assist in devising strategies for improvement of the overall system.

3.3. Flood vulnerability index methodology

The Flood Vulnerability Index (FVI) aims to identify hotspots related to flood risk in different regions of the world, so that it can be applied as a tool to assist planners and policy makers in prioritising their areas of intervention and also as an instrument to provide useful information for awareness raising. The main concept consists of identifying different characteristics of a system, making it applicable to floods on different spatial levels.

Connor & Hiroki (2005) presented a methodology to calculate a FVI for river basins, using eleven indicators divided in four components. The index uses two sub-indices for its computation; the human index, which corresponds to the social effects of floods, and the material index, which covers the economic effects of floods.

This chapter describes a revised methodology to compute a FVI, based on indicators, aimed at assessing the conditions which induce flood damage at various spatial scales. The methodology, in principle, is based on sets of indicators for the four different factors of vulnerability for fluvial and urban floods.

The methodology recognises different characteristics for each spatial scale identified, allowing a more in-depth analysis and interpretation of local indicators. It also allows selection of actions to diminish local flood vulnerability. The whole concept of FVI is that we have a hazard, in this case a flood event, which is affecting the system (river basin, sub-catchment or urban area) in four of its main components (social, economic, environmental and physical). This system is exposed and susceptible to floods, but also has its own resilience.

3.3.1 Selection of the vulnerability indicators

Vulnerability indicators are commonly used in vulnerability assessment. The first step in an indicator-based vulnerability assessment is to select indicators. The standard practice is to assemble a list of indicators using criteria such as: suitability, following a conceptual framework or definitions and availability of data.

Vulnerability needs to be reflected through the indicators chosen. The indicators should allow decision and policy makers to recognize and set goals and provide guidance for strategies to reduce vulnerability. The vulnerability indicators should provide additional information to set more precise and quantitative targets for vulnerability reduction. System indicators facilitate the analysis of the relative state of the overall system and they should reflect the socio-economic, environmental and physical condition of the geographic region.

As mentioned in Section 2.1.1. the procedures for indicator selection follow two general approaches, a deductive research approach and the second inductive research approach.

Since the development of the FVI involves the understanding of different relational situations and characteristics of a system with flood events, a deductive approach to identify the best possible indicators has been used, based on existing principles and the conceptual framework (dividing flood vulnerability indicators among flood vulnerability factors and vulnerability components). Understanding the causes of floods and their main effects on the different components of a system led to the recognition of the optimal indicators (Figure 3.1).

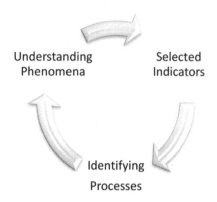

Figure 3.1 Deductive approach processes

Almost 80 potential indicators have been examined to upgrade the existing methodology of Connor and Hiroki (2005), taking into account the previously mentioned geographical scales: that is, river basin (R), sub-catchment (S) and urban areas (U). A composite index approach was used, as also used in the construction of Climate Vulnerability Index (Sullivan & Meigh, 2005).

3.3.2. General FVI equation for all scales

The proposed general FVI Equation (3.1) links the values of all indicators to flood vulnerability components and factors, without balancing or interpolating from a series of data. Using this equation allows comparisons between different geographical scales, since the outcome of the computation is dimensionless. Dimensionless results are necessary in order to compare FVI's for similar components and scales for different case studies.

Dimensionless FVI equations are developed by using fractions with indicators as part of a numerator or denominator, depending on their effect on flood vulnerability. Indicators representing exposure and susceptibility increase the flood vulnerability and are therefore placed in the nominator. The resilience indicators decrease flood vulnerability and are conversely part of the denominator:

$$FVI = \frac{E * S}{R}$$
3.1

Besides the FVI values for each component, standardised results are developed for further comparison between components (Sullivan et al., 2003), also serving the purpose of easier interpretation. Standardised FVI values range between 1 and 0; 1 being the most vulnerable to floods. The standardised formula is presented as a FVI of a system divided by the maximum FVI within one system (3.2):

$$sFVI = \frac{FVI_{specific}}{\max_{i=1}^{n}(FVI)_i}$$
3.2

Since the study of river basins covers large heterogeneous areas, interpreting the FVI on such a scale can be misleading. Therefore the study of smaller spatial scales can lead to a more accurate evaluation of the flood vulnerability of a region (Balica et al., 2009). Interpreting the values of all sub-catchments in one river basin can provide a more detailed image of the situation in the basin.

The relation of vulnerability components, indicators and factors is illustrated in Table 3.1 for various spatial scales. The availability of data, the importance of certain indicators and the condition that all FVI's computed must be dimensionless for the purposes of comparison, led to the formulation of the equations for each scale and for each vulnerability component.

The importance of selecting indicators is real, the FVI can be changed by, for example, decreasing the protection of nature areas, or increasing that of cities, while flood impacts may be reduced by e.g. raising flood risk awareness and changing land use. These measures may increase the resilience of the system as a whole, since expected damages are lowered, recovery is enhanced and the reaction to flood may become more gradual.

Table 3.1. Relationship between components, indicators and factors

<table>
<tr><th colspan="10">Overall Indicators</th></tr>
<tr><th colspan="10">Relationship between components and factors</th></tr>
<tr><th>Flood Vulnerability</th><th colspan="3">Exposure</th><th colspan="3">Susceptibility</th><th colspan="3">Resilience</th></tr>
<tr><th></th><th></th><th>Abb</th><th>Geographic Scale</th><th></th><th>Abb</th><th>Geographic Scale</th><th></th><th>Abb</th><th>Geographic Scale</th></tr>
<tr><td rowspan="13">Social Component</td><td>Population Density</td><td>Pd</td><td>R,S,U</td><td>Past Experience</td><td>PE</td><td>R,S,U</td><td>Warning system</td><td>WS</td><td>R,S,U</td></tr>
<tr><td>Population in Flood area</td><td>Pfa</td><td>R,S,U</td><td>Education (Literacy rate)</td><td>Ed</td><td>R,S,U</td><td>Evacuation Routes</td><td>ER</td><td>R,S,U</td></tr>
<tr><td>Closeness to inundation area</td><td>Cia</td><td>R,S,U</td><td>Preparedness/Awareness</td><td>A/P</td><td>R,S,U</td><td>Institutional Capacity</td><td>IC</td><td>R,S,U</td></tr>
<tr><td>Population close to coastal</td><td>Pcol</td><td>R,S,U</td><td>Child Mortality</td><td>Cm</td><td>R,S,U</td><td>Emergency service</td><td>ES</td><td>R,S,U</td></tr>
<tr><td>Population under poverty</td><td>Pp</td><td>R,S,U</td><td>Communication penetration rate</td><td>CPR</td><td>R,S,U</td><td>Shelters</td><td>S</td><td>R,S,U</td></tr>
<tr><td>% of Urbanized area</td><td>%UA</td><td>R,S</td><td>Population with access to sanitation</td><td>PwaS</td><td>R,S,U</td><td></td><td></td><td></td></tr>
<tr><td>Rural population</td><td>Rpop</td><td>R,S</td><td>Rural population w/o access to WS</td><td>PwoWS</td><td>R,S</td><td></td><td></td><td></td></tr>
<tr><td>Cadastre Survey</td><td>CS</td><td>S,U</td><td>Quality of Water Supply</td><td>QWS</td><td>S,U</td><td></td><td></td><td></td></tr>
<tr><td>Cultural Heritage</td><td>CH</td><td>S,U</td><td>Quality of Energy Supply</td><td>QES</td><td>S,U</td><td></td><td></td><td></td></tr>
<tr><td>% of disable</td><td>%disable</td><td>U</td><td>Population Growth</td><td>PG</td><td>S,U</td><td></td><td></td><td></td></tr>
<tr><td></td><td></td><td></td><td>Human Health</td><td>HH</td><td>S,U</td><td></td><td></td><td></td></tr>
<tr><td></td><td></td><td></td><td>Human Development Index</td><td>HDI</td><td>S,U</td><td></td><td></td><td></td></tr>
<tr><td></td><td></td><td></td><td>Urban Planning</td><td>UP</td><td>U</td><td></td><td></td><td></td></tr>
<tr><td rowspan="9">Economic Component</td><td>Land Use</td><td>LU</td><td>R,S,U</td><td>Unemployment</td><td>UM</td><td>R,S,U</td><td>Investment in c. measure</td><td>AmIn</td><td>R,S,U</td></tr>
<tr><td>Proximity to river</td><td>PR</td><td>R,S,U</td><td>Income</td><td>I</td><td>R,S,U</td><td>Infrastructure management</td><td>IM</td><td>R,S,U</td></tr>
<tr><td>Closeness to inundation area</td><td>Cia</td><td>R,S,U</td><td>Inequality</td><td>Ineq</td><td>R,S,U</td><td>Dams & Storage Capacity</td><td>DSC</td><td>R,S,U</td></tr>
<tr><td>% of Urbanized area</td><td>%UA</td><td>R,S</td><td>Yearly Volume</td><td>Vyear</td><td>R,S,U</td><td>Flood Insurance</td><td>FI</td><td>R,S,U</td></tr>
<tr><td>Cadastre Survey</td><td>CS</td><td>S,U</td><td>Life Expectancy Index</td><td>LEI</td><td>R,S,U</td><td>Economic Recovery</td><td>ECR</td><td>R,S,U</td></tr>
<tr><td></td><td></td><td></td><td>Urban Growth</td><td>UG</td><td>S,U</td><td>Past experience</td><td>PE</td><td>S,U</td></tr>
<tr><td></td><td></td><td></td><td>Child Mortality</td><td>CM</td><td>S,U</td><td>Dikes/levees</td><td>DL</td><td>S,U</td></tr>
<tr><td></td><td></td><td></td><td>Regional GDP/capita</td><td>GDP</td><td>S</td><td></td><td></td><td></td></tr>
<tr><td></td><td></td><td></td><td>Urban Planning</td><td>UP</td><td>U</td><td></td><td></td><td></td></tr>
<tr><td rowspan="8">Environmental Component</td><td>Ground WL</td><td>GWL</td><td>R,S,U</td><td>Natural Reservations</td><td>NR</td><td>R,S,U</td><td>Recovery time to floods</td><td>RTF</td><td>R,S,U</td></tr>
<tr><td>Land Use</td><td>LU</td><td>R,S,U</td><td>Years of sustaining health life</td><td>YSHL</td><td>R,S,U</td><td>Environmental concern</td><td>EC</td><td>R,S,U</td></tr>
<tr><td>Over used area</td><td>OUA</td><td>R,S,U</td><td>Quality of infrastructure</td><td>QI</td><td>R,S,U</td><td></td><td></td><td></td></tr>
<tr><td>Degraded area</td><td>DA</td><td>R,S,U</td><td>Human health</td><td>HH</td><td>S,U</td><td></td><td></td><td></td></tr>
<tr><td>Unpopulated land area</td><td>Unpop</td><td>R,S</td><td>Urban growth</td><td>UG</td><td>S,U</td><td></td><td></td><td></td></tr>
<tr><td>Types of vegetation</td><td>TV</td><td>R,S</td><td>Child Mortality</td><td>CM</td><td>S,U</td><td></td><td></td><td></td></tr>
<tr><td>% of Urbanized area</td><td>%UA</td><td>R,S</td><td>Rainfall</td><td>Rainfall</td><td></td><td></td><td></td><td></td></tr>
<tr><td>Forest change rate</td><td>FCR</td><td>R</td><td>Evaporation</td><td>Ev</td><td></td><td></td><td></td><td></td></tr>
<tr><td rowspan="14">Physical Component</td><td>Topography (slope)</td><td>T</td><td>R,S,U</td><td>Buildings Codes</td><td>Bc</td><td>U</td><td>Dams & Storage Capacity</td><td>DSC</td><td>R,S,U</td></tr>
<tr><td>Heavy rainfall</td><td>HR</td><td>R,S,U</td><td>Frequency of Occurance</td><td>FO</td><td>R,S,U</td><td>Roads</td><td>R</td><td>R,S,U</td></tr>
<tr><td>Flood Duration</td><td>FD</td><td>R,S,U</td><td></td><td></td><td></td><td>Dikes/Levees</td><td>DL</td><td>S,U</td></tr>
<tr><td>Return Periods</td><td>RP</td><td>R,S,U</td><td></td><td></td><td></td><td></td><td></td><td></td></tr>
<tr><td>Proximity to river</td><td>PR</td><td>R,S,U</td><td></td><td></td><td></td><td></td><td></td><td></td></tr>
<tr><td>Soil Moisture</td><td>SM</td><td>R,S,U</td><td></td><td></td><td></td><td></td><td></td><td></td></tr>
<tr><td>Evaporation Rate</td><td>Ev</td><td>R,S,U</td><td></td><td></td><td></td><td></td><td></td><td></td></tr>
<tr><td>River Discharge</td><td>RD</td><td>R,S,U</td><td></td><td></td><td></td><td></td><td></td><td></td></tr>
<tr><td>Flow Velocity</td><td>FV</td><td>S,U</td><td></td><td></td><td></td><td></td><td></td><td></td></tr>
<tr><td>Storm Surge</td><td>SS</td><td>S,U</td><td></td><td></td><td></td><td></td><td></td><td></td></tr>
<tr><td>Rainfall</td><td>Rainfall</td><td>S,U</td><td></td><td></td><td></td><td></td><td></td><td></td></tr>
<tr><td>Flood Water Depth</td><td>FWD</td><td>S,U</td><td></td><td></td><td></td><td></td><td></td><td></td></tr>
<tr><td>Sedimentation Load</td><td>SL</td><td>S,U</td><td></td><td></td><td></td><td></td><td></td><td></td></tr>
<tr><td>Yearly Volume</td><td>Vyear</td><td>S,U</td><td></td><td></td><td></td><td></td><td></td><td></td></tr>
</table>

3.3.3. Flood Vulnerability Index at river basin scale

A river basin is the portion of land drained by a river and its tributaries. It encompasses the entire land surface dissected and drained by many streams and creeks that flow downhill into one another, and eventually into one river. The final destination is a lake, an estuary or an ocean.

In general river basins require information from more than one country, therefore sub-catchments and urban areas have to be considered and represented as a system in their own. The data of each country must be interpolated to reflect the reality of the area of study and not of the entire country.

The river basin is the largest scale studied for this thesis. It may include river basins as big as the Amazon River, the largest in the world with more than 7,000,000 km^2, or as small as Rhine River, 185,000 km^2, or Tagus River 81,600 km^2.

In total 58 indicators have been taken into consideration for this geographical scale. However 26 indicators were used to develop the equations for the river basin FVI's, for each flood vulnerability factor and component. The remaining indicators were not applied because of difficulties in developing a dimensionless FVI, redundancy of definitions or complexity of obtaining the data (See Appendix 3.1).

3.3.3.1. Equations of the river basin scale

The equations presented for vulnerability components at the river basin scale, show the indicators as a ratio, favouring the omission of units. Each FVI component has its own range of values, depending on the numerical values of the indicators, reflecting the need to evaluate each component on its own.

On a global perspective the results will be presented in values between 0 and 1; 1 being the highest vulnerability found in the samples studied and 0 the lowest vulnerability. This procedure will be used for all geographical scales, taking care that comparisons will be done only on merits of higher relative vulnerability within the sample.

Flood Vulnerability Index for social component on river basin scale:

$$FVI_S = \left[\frac{P_{FA.}\,C_{M.}\,Um}{P_{E.}\,AP, C_{PR}, HDI, W_S, E_R} \right] \qquad 3.3$$

$$Dimension\_of\_FVI_S = \frac{[persons][\%][\%]}{[persons][-][\%][-][-][\%]} \text{ - dimensionless;}$$

Flood Vulnerability Index for economic component on river basin scale:

$$FVI_{EC} = \left[\frac{L_U.\,Um, I_{neq}, HDI}{AmInv, E_R, S_C / YearDischarge} \right] \qquad 3.4$$

$$Dimension\_of\_FVI_{Ec} = \frac{[\%][\%][-][\%]}{[-][-][m^3/m^3]} \text{ - dimensionless;}$$

Flood Vulnerability Index for environmental component on river basin scale:

$$FVI_{En} = \left[\frac{R_{a\,inf\,all.}D_A}{N_R, E_V, U_{npop}, L_U} \right] \qquad 3.5$$

$$Dimension\_of\_FVI_{En} = \frac{[m/year][\%]}{[\%][m/year][\%][\%]} \text{ - dimensionless;}$$

Flood Vulnerability Index for physical component on river basin scale:

$$FVI_{Ph} = \left[\frac{T, D_{HR}, R_D, F_o}{E_V / R_{a\,inf\,all}}, D\_S_C \right] \qquad 3.6$$

$$Dimension\_of\_FVI_{Ph} = \frac{[-][\#][m^3/s][year]}{\dfrac{mm/year}{mm/year}[m^3]} *86400*365$$

3.3.4. Flood Vulnerability Index at sub-catchment scale

The term sub-catchment describes an area of land that drains part of a river basin down slope to the lowest point. The water moves through a network of drainage pathways, underground and on the surface. Generally, these pathways converge into streams and rivers, which become progressively larger as the water moves on downstream, eventually reaching an estuary and the ocean. Other terms used interchangeably with watershed include drainage basin or catchment basin.

The FVI methodology for the sub-catchment scale was initially developed by using a total of 71 indicators. Since the development of the FVI involves the understanding of different relational situations and characteristics of a system with flood events, a deductive approach to identify the best possible indicators has been used. Understanding the causes of floods and their main effects on the different components of a system led to the recognition of the optimal indicators.

However, only 28 indicators have been selected for the sub-catchment FVI equations (See Appendix 3.2).

3.3.4.1. Equations of the sub-catchment scale

Equations (3.8) to (3.11) reflect the vulnerability of a selected geographical area, limited by watershed divisions rather than administrative boundaries. The latter often adds to the difficulty of collecting data.

Flood Vulnerability Index for social component on sub-catchment scale:

$$FVI_S = FVI_S \left[\frac{P_{FA}, R_{Pop}, \%_{disable}, C_m}{P_E, A/P, C_{PR}, W_S, E_R, HDI} \right]$$

3.8

Dimension of $FVI_S = \dfrac{[persons][\%][\%][-]}{[persons][-][\%][-][\%][-]}$ - dimensionless;

Flood Vulnerability Index for economic component on sub-catchment scale:

$$FVI_{Ec} = FVI_{Ec} \left[\frac{L_U, U_M, I_{neq}, U_A}{L_{EI}, F_I, AmInv, {}^{S_C}\!/_{Vyear}, E_{CR}} \right]$$

3.9

Dimension of $FVI_{Ec} = \dfrac{[\%][\%][-][\%]}{[-][-][euro/euro][m^3/m^3][-]}$ - dimensionless;

Flood Vulnerability Index for environmental component on sub-catchment scale:

$$FVI_{En} = FVI_{En} \left[\frac{R_{a\,inf\,all}, D_A, U_G}{L_U, E_V, N_R, U_{npop}} \right]$$

3.10

Dimension of $FVI_{En} = \dfrac{[\%][\%][m/year]}{[\%][m/year][\%][\%]}$ - dimensionless;

Flood Vulnerability Index for physical component on sub-catchment scale:

$$FVI_{Ph} = FVI_{Ph} \left[\frac{T}{{}^{E_V}\!/_{R_{a\,inf\,all}}, {}^{S_C}\!/_{Vyear}, D\_L} \right]$$

3.11

Dimension of $FVI_{Ph} = \dfrac{[-]}{{}^{[mm/year]}\!/_{[mm/year]}, {}^{[m^3}\!/_{m^3]}, {}^{[Km}\!/_{Km]}}$ - dimensionless;

3.3.5. Flood Vulnerability Index at urban area scale

The usual concept of a town, urban area, would be "a free-standing built-up area with a service core with a sufficient number and variety of shops and services, including a market" (Statistics UK, 2001). An urban area would have administrative, commercial, educational, entertainment and other social and civic functions and, evidence of being historically well established, as well as a local network of roads and other means of transport would focus on the area, and it would be a place drawing people for services and employment from surrounding areas.

The urbanisation process itself is one of the causes of flood disasters. The loss of natural retention areas, previously provided by marsh paddy and other agricultural areas, due to urban expansion has allowed floodwater to travel more quickly to receiving streams, swelling them beyond their capacity (UNU, 2005). The phenomenon is exacerbated by the paved urban landscape and the continuing urbanisation. Adding that the urban areas are highly dense populated make them especially vulnerable to flood effects.

63 indicators have been considered for this geographical scale (See Appendix 3.3).

3.3.5.1. Equations of the urban area scale

Flood vulnerability index for social component on urban area scale:

$$FVI_S = \left[\frac{P_{FA}, R_{Pop}, \%disable, Cm}{P_E, A/P, C_{PR}, W_S, E_R, HDI} \right] \qquad 3.12$$

$$Dimension\_of\_FVI_S = \frac{[persons][\%][\%][-]}{[persons][-][\%][-][\%][-]} \text{ - dimensionless}$$

Flood vulnerability index for economic component on urban area scale:

$$FVI_{Ec} = \left[\frac{L_U, U_m, I_{neq}, U_A}{L_{EI}, F_I, AmInv, {S_C}/{Vyear}, E_{CR}} \right] \qquad 3.13$$

$$Dimension\_of\_FVI_{Ec} = \frac{[\%][\%][-]\%}{[-][-][euro/{euro}]{m^3}/{m^3}[-]} \text{ - dimensionless}$$

Flood vulnerability index for environmental component on urban area scale:

$$FVI_{En} = \left[\frac{U_G, R_{a\,inf\,all}}{E_V, L_U} \right] \qquad 3.14$$

$$Dimension\_of\_FVI_{En} = \frac{[\%]\left[\dfrac{m}{year}\right]}{\left[\dfrac{m}{year}\right][\%]} \text{ - dimensionless}$$

Flood vulnerability index for physical component on urban area scale:

$$FVI_{Ph} = \left[\frac{T, C_R}{E_V\Big/R_{a\,inf\,all}, \dfrac{S_C}{Vyear}, D\_L}\right] \qquad 3.15$$

$$Dimension\_Of\_FVI_{Ph} = \frac{[-][km]}{\left[\dfrac{mm}{year}\right]\Big/\left[\dfrac{mm}{year}\right]\Big/\dfrac{m^3}{m^3}, [km]} \text{ - dimensionless}$$

This revised methodology was applied to three river basins Danube, Rhine and Mekong, five sub-catchments Tisza (Hungry), Timis (Romania), Bega (Romania), Mun (Cambodia) and Neckar (Germany) and three urban areas Timisoara (Romania), Mannheim (Germany) and Phnom Penh (Cambodia), the results of this application can be found on the FVI network of knowledge, described in the Section below.

3.4. The FVI network of knowledge – implementation and dissemination

The aim of a FVI web-site (unesco-ihe-fvi.org) is to create a network of knowledge among different institutions in which this methodology is applied. It is also hoped to encourage collaboration between the members of the network on managing flood vulnerability information and also promote further studies on flood risk assessment at all scales. The FVI website can be very useful in dealing with uncertainty, because it can identify areas that are especially vulnerable and which require priority measures. Many of the measures that can be taken based on the results of the FVI are useful even if no climate change impacts will occur (i.e. low-regret or no-regret measures).

The network of knowledge fundamental concept consists of identifying, from various indicators, the different characteristics of a system that will make it vulnerable to floods on different levels.

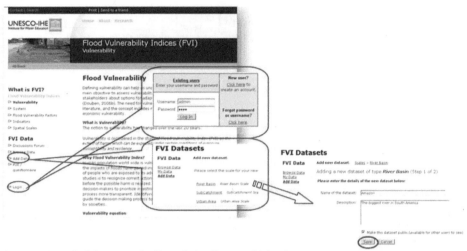

Figure 3.2. Detail of the network of knowledge (log in, add data)

Flood vulnerability assessment can be an important component of a response to uncertainty, as part of either a bottom-up approach such as resilience (Dessai and Hulme, 2007) or a top-down approach such as risk assessment. For instance, FVI can help policy makers or governments to determine how limited resources could be used in other to reap maximum benefits in terms of flood protection and crisis management. The information obtained can enable decision makers to be better prepared in the face of uncertainties and in the event of a crisis.

Vulnerability assessment is also important for a human development approach. Thus, the FVI can be a measure to reduce the negative impacts of uncertainty and to better prepare the most vulnerable and potentially most affected areas to deal with an uncertain climatic future.

The network of knowledge web interface contains a collaboration room for collecting data for each spatial scale along with information and knowledge concerning the concept of vulnerability (unesco-ihe-fvi.org). Any user can create an account on the website and can log on to add their data, shown in Figure 3.2 (private or shared). This predefined and flexible form contains the indicators used in the FVI equations for each spatial scale. The indicators list can be continuously improved based on community feedback and future evolution. After entering the data, fast computations are carried out to calculate the vulnerability of each area and to store the results.

All data are stored in a relational database using a flexible table model for easier extension and improvement (adding indicators, descriptions, etc.) seen in Figure 3.3. The results are stored in the database tables and can be displayed when browsing the data. As said before, one major strength of this index is that it combines many different components (social, economic, environmental and physical) (Equation 3.16) and three different factors of vulnerability (exposure, susceptibility and resilience). The different results available on the website can be then easily interpreted and compared (Figures 3.4 and 3.5).

$$\text{Total FVI} = {}_{\text{social}}\text{FVI} + {}_{\text{economic}}\text{FVI} + {}_{\text{Environmental}}\text{FVI} + {}_{\text{physical}}\text{FVI} \qquad 3.16$$

The main advantage of the FVI is that it can be used as a rational exercise on each component of flood vulnerability in a particular region, in order to determine possible ways to increase the resilience of the analysed region.

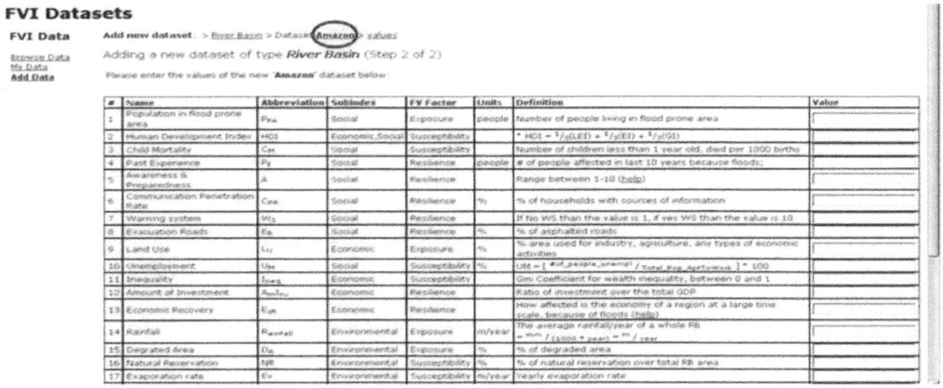

FVI Datasets

FVI Data

Add new dataset: > River Basin > Dataset Amazon > values

Adding a new dataset of type *River Basin* (Step 2 of 2)

Please enter the values of the new 'Amazon' dataset below:

Browse Data
My Data
Add Data

#	Name	Abbreviation	Subindex	FV Factor	Units	Definition	Value
1	Population in flood prone area	P_{FA}	Social	Exposure	people	Number of people living in flood prone area	
2	Human Development Index	HDI	Economic,Social	Susceptibility		* HDI = 1/3(LEI) + 1/3(EI) + 1/3(GI)	
3	Child Mortality	C_M	Social	Susceptibility		Number of children less than 1 year old, died per 1000 births	
4	Past Experience	P_E	Social	Resilience	people	# of people affected in last 10 years because floods;	
5	Awareness & Preparedness	A	Social	Resilience		Range between 1-10 (help)	
6	Communication Penetration Rate	C_{PR}	Social	Resilience	%	% of households with sources of information	
7	Warning system	W_S	Social	Resilience		If No WS than the value is 1, if yes WS than the value is 10	
8	Evacuation Roads	E_R	Social	Resilience	%	% of asphalted roads	
9	Land Use	L_U	Economic	Exposure	%	% area used for industry, agriculture, any types of economic activities	
10	Unemployment	U_M	Social	Susceptibility	%	UM = [#of_people_unempl / Total_Pop_AgeToWork] * 100	
11	Inequality	I_{neq}	Economic	Susceptibility		Gini Coefficient for wealth inequality, between 0 and 1	
12	Amount of Investment	A_{mInv}	Economic	Resilience		Ratio of investment over the total GDP	
13	Economic Recovery	E_{cR}	Economic	Resilience		How affected is the economy of a region at a large time scale, because of floods (help)	
14	Rainfall	$R_{ainfall}$	Environmental	Exposure	m/year	The average rainfall/year of a whole RB = mm / (1000 * year) = m / year	
15	Degrated Area	D_A	Environmental	Exposure	%	% of degraded area	
16	Natural Reservation	NR	Environmental	Susceptibility	%	% of natural reservation over total RB area	
17	Evaporation rate	Ev	Environmental	Susceptibility	m/year	Yearly evaporation rate	

Figure 3.3. Feature of table of indicators

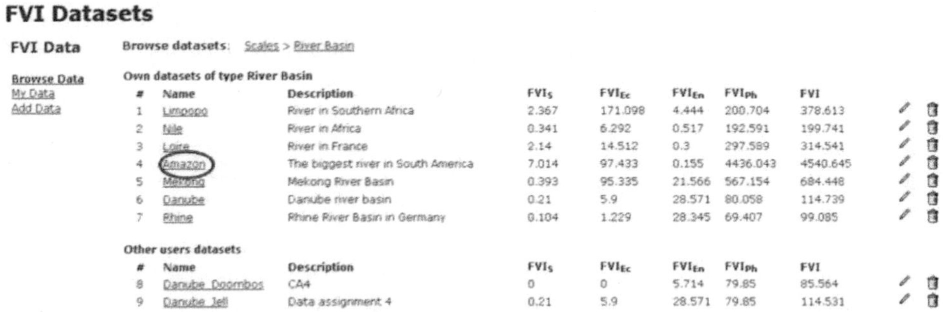

FVI Datasets

FVI Data

Browse datasets: Scales > River Basin

Browse Data
My Data
Add Data

Own datasets of type River Basin

#	Name	Description	FVI_S	FVI_{Ec}	FVI_{En}	FVI_{Ph}	FVI
1	Limpopo	River in Southern Africa	2.367	171.098	4.444	200.704	378.613
2	Nile	River in Africa	0.341	6.292	0.517	192.591	199.741
3	Loire	River in France	2.14	14.512	0.3	297.589	314.541
4	Amazon	The biggest river in South America	7.014	97.433	0.155	4436.043	4540.645
5	Mekong	Mekong River Basin	0.393	95.335	21.566	567.154	684.448
6	Danube	Danube river basin	0.21	5.9	28.571	80.058	114.739
7	Rhine	Rhine River Basin in Germany	0.104	1.229	28.345	69.407	99.085

Other users datasets

#	Name	Description	FVI_S	FVI_{Ec}	FVI_{En}	FVI_{Ph}	FVI
8	Danube_Doornbos	CA4	0	0	5.714	79.85	85.564
9	Danube_Jell	Data assignment 4	0.21	5.9	28.571	79.85	114.531

Figure 3.4. Screenshot of the list of FVI case studies results

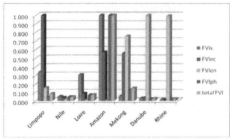 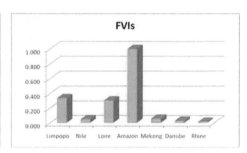

Figure 3.5. Graphics of the FVI case studies results

Used by local stakeholders, it can permit discussion on what component requires most attention and how to improve risk management, particularly in the pro-action and prevention steps.

3.5. Discussions

The methodology presented here is based on sets of indicators for different factors and different geographical scales, focusing on fluvial and urban floods.

Various indicators have been taken into account to assess flood vulnerability. Some of the indicators originally proposed were not considered in the final equations, due to the difficulty of quantifying them, finding data, possible redundancy with other indicators or with the purpose of creating a dimensionless result for each equation. Since the methodology is based on indicators, its main weakness is the accuracy of the data. For the results to be valid, all data must be derived from reliable sources, specified for a precise spatial area at a defined time.

The FVI can be used in combination with other decision-making tools, and specifically include participatory methods with the population of areas identified as vulnerable, a team of multidisciplinary thematic specialists and representatives of the society and those with expert judgment.

Regarding the network of knowledge, the disadvantage of a collaborative tool is that invalid datasets (tests) can be erroneously entered by a user. This can be overcome by using a validation process involving the user and the administrator. Thus, at the end of the validation process, the administrator can flag a dataset as being validated/checked and users can filter only valid datasets if they wish.

Through this network of knowledge, the vulnerability index can help to assess and to improve the links of the safety chain in risk management. The indicators used by the FVI can thus help to analyse, at any moment in time (based on updated information/values), the actual risk and the preparedness for flooding and contribute to adequate planning of measures to limit the risk and the vulnerability within the safety chain.

3.6. Conclusions

The conclusions concerning the development and the network of knowledge of a FVI methodology can be summarised as follows:

- The FVI is applicable on three different spatial scales: river basin, sub-catchment and urban areas;
- FVI provides a method to systematically express the vulnerability of a river system to disruption factors, such as floods;
- FVI offers easy to understand results, with the use of a single value to characterize high or low vulnerability. This also allows continuous data interpretation for more in-depth analysis and it is suitable to policy-makers;
- The use of the FVI methodology improves the decision-making process by identifying the vulnerability of flood prone areas;
- The FVI is a powerful tool for policy and decision makers to prioritise investments and makes the decision-making process more transparent. Identifying areas with high flood vulnerability may guide the decision-making process towards better means of dealing with floods;
- It is believed that the FVI website will be very useful in developing the index further and in developing a network of researchers and practitioners in this field globally. Over time, different methodologies can be developed and incorporated into the website. The tool will present a means of assessing vulnerability in a future that is uncertain.

Appendices

Appendix 3.1. River basin scale indicators

No	Abb.	Name	Sub-index	FV Factor	Units	Definition of indicator	Functional relationship with vulnerability	Data Source
						River Basin Scale		
1	P_{FA}	Population in flood prone area	FVI_S	E	people	Number of people living in flood prone area	The higher number of people, higher vulnerability	CRED
2	HDI	Human Development Index	$FVI_{S,Ec}$	S	-	*HDI $= \frac{1}{3}(LEI) + \frac{1}{3}(EI) + \frac{1}{3}(GI)$	The higher value, lower vulnerability	UNDP, 2004
3	C_M	Child Mortality	FVI_S	S	-	Number of children less than 1 year old, died per 1000 births	The higher number of chilren, higher vulnerability	EPI Report,2006
4	P_E	Past Experience	FVI_S	R	people	# of people affected in last 10 years because floods;	The higher value, lower vulnerability	EM-DAT
5	A	Awareness&Preparedness	FVI_S	R	-	Range between 1-10	10 means lower vulnerability	Refer to Table
6	C_{PR}	Communication Penetration Rate	FVI_S	R	%	% of households with sources of information	Higher percentage means lower vulnerability	INTUTE
7	W_S	Warning system	FVI_S	R	-	if No W_S than the value is 1, if yes W_S than the value is 10	Having WS reduces the vulnerability	Yes/No
8	E_R	Evacuation Roads	FVI_S	R	%	% of asphalted roads.	The better the quality of roads, improves the evacuation during floods	INTUTE
9	L_U	Land Use	FVI_{Ec}	E	%	% area used for industry, agriculture, any types of economic activities	The higher %, the high vulnerability	WRI
10	U_M	Unemployment	$FVI_{S,ec}$	S	%	$U_M = \frac{\#of\_people\ Unempl}{Total\_Pop\_AptToWork} * 100$	The higher %, the high vulnerability	World Factbook 2005
11	I_{neq}	Inequality	FVI_{Ec}	S	-	Gini Coefficient for wealth inequality, between 0 and 1	Where 1 means low vulnerability	UN
12	A_{mInv}	Amount of Investment	FVI_{Ec}	R	-	Ratio of investment over the total GDP	Higher the investment lower vulnerability	MRC,2006 IRMA project
13	Er	Economic Recovery	FVI_{Ec}	R	-	How affected is the economy of a region at a large time scale, because of floods	Higher the recovery lower vulnerability	Refer to Table
14	$R_{ainfall}$	Rainfall	FVI_{En}	E	m/year	the average rainfall/year of a whole RB $R_{ainfall} = \frac{mm}{1000 * year} = \frac{m}{year}$	Higher rainfall, higher vulnerability	MRC,2006; Ekstrom et al.
15	D_A	Degrated Area	FVI_{En}	E	%	% of degraded area	Bigger D_A, higher vulnerability	WRI
16	N_R	Natural Reservation	FVI_{En}	S	%	% of natural reservation over total RB $N_R = \frac{A_{NR}}{Total\_Area\_of\_River\_Basi} * 100$	Higher %, Lower vulnerability	WRI
17	Ev	Evaporation Rate	$FVI_{En,Ph}$	E	m/year	yearly evaporation rate	higher GWL, higher vulnerability	Ekstrom et al. MRC
18	U_{npop}	Unpopulated Area	FVI_{En}	E	%	% of area with density of population less than 10 pers/km^2	Higher Unpop area. Lower vulnerability	WRA
19	L_U	Land Use	FVI_{En}	E	%	% of forested area	The higher %, the low vulnerability	WRI
20	T	Topography	FVI_{Ph}	E	-	average slope of river basin	The steeper slope, higher vulnerability	FVI
21	D_{HR}	# of days with heavy rainfall	FVI_{Ph}	E	#	number of days with heavy rainfall, more than 100mm/day	higher # of days, higher vulnerability	FVI
22	R_O	River Discharge	FVI_{Ph}	E	m^3/s	maximum discharge in record of the last 10 years, m^3/s	higher RD, higher vulnerability	ICPDR, UNH/GRDC
23	Fo	Frequency of occurance	FVI_{Ph}	E	years	years between floods	bigger # of years, high vulnerability	
24	$Ev/R_{ainfall}$	Evaporation rate/Rainfall	FVI_{Ph}	E	-	Yearly Evaporation over yearly rainfall	Higher the Ev, lower vulnerability	MRC, Ekstrom et al.
25	D_Sc	Dams_Storage capacity	FVI_{Ph}	R	m^3	The total volume of water, which can be stored by dams, polders, etc.	higher m3, higher vulnerability	WB
26	Sc/D	Storage capacity over yearly discharge	FVI_{Ec}	R	m3/m3	Storage capacity divided by yearly volume runoff	higher Sc means lower vulnerability	reffer to 25 and average discharge

Appendix 3.2. Sub-catchment scale Indicators

No	Abb.	Name	Sub-index	FV Factor	Units	Definition of indicator	Functional relationship with vulnerability	Data Source
						Sub-catchment Scale		
1	PD	Population density	FVI$_S$	E	people/km^2	There is an important exposure to a given hazard if population is concentrated	Higher # of people, higher vulnerability	http://www.dams.org/docs/kbase/studies /csthanx.pdf
2	P$_{FA}$	Population in flood prone area	FVI$_S$	E	people	Number of people living in flood prone area	The higher number of people, higher vulnerability	http://www.dams.org/docs/kbase/studies /csthanx.pdf
3	U$_A$	Urbanized Area	FVI$_{S,Ec}$	E	%	% of total area which is urbanized	higher %,higher vulnerability	INTUTE, PELCOM
4	R$_{pop}$	Rural population	FVI$_S$	E	%	% of population living outside of urbanized area	higher %,higher vulnerability	http://www.dams.org/docs/kbase/studies /csthanx.pdf
5	% of disable	Disable People	FVI$_S$	E	%	% of population with any kind of disabilities, also people less 12 and more than 65	higher %,higher vulnerability	INTUTE
6	HDI	Human Development Index	FVI$_S$	S	-	*HDI $= \frac{1}{3}(LEI) + \frac{1}{3}(EI) + \frac{1}{3}(GI)$	The higher value, lower vulnerability	UNDP, 2004
7	C$_M$	Child Mortality	FVI$_S$	S	-	Number of children less than 1 year old, died per 1000 births	The higher number of chilren, higher vulnerability	http://www.dams.org/docs/kbase/studies /csthanx.pdf, EPI
8	P$_E$	Past Experience	FVI$_S$	R	people	# of people who have been affected in last 10 years because flood events;	The higher value, lower vulnerability	http://69.178.233.117/MPcomp/2003/ma ps/Academicindividual-Caquard.pdf, Selected Global Extreme Information, Reuter News, EM-DAT
9	A	Awareness&Preparedness	FVI$_S$	R	-	Range between 1-10	10 means lower vulnerability	reffer to table
10	C$_{PR}$	Communication Penetration Rate	FVI$_S$	R	%	% of households with sources of information	Higher percentage means lower vulnerability	INTUTE
11	W$_S$	Warning system	FVI$_S$	R	-	if No W$_S$ than the value is 1, if yes W$_S$ than the value is 10	Having WS reduces the vulnerability	Y/N
12	E$_R$	Evacuation Roads	FVI$_S$	R	%	% of asphalted roads.	The better the quality of roads, improves the evacuation during floods	INTUTE
13	L$_U$	Land Use	FVI$_{Ec}$	E	%	% area used for industry, agriculture, any types of economic activities	The higher %, the high vulnerability	INTUTE
14	U$_M$	Unemployment	FVI$_{Ec}$	S	%	$\frac{\#of\_people\_Unempl}{Total\_Pop\_AptToWork} * 100$	The higher %, the high vulnerability	World Factbook 2005
15	I$_{neq}$	Inequality	FVI$_{Ec}$	S	-	Gini Coefficient for wealth inequality, between 0 and 1	Where 1 means low vulnerability	UN
16	L$_{EI}$	Life expectancy	FVI$_{Ec}$	S	-	LEI= $\frac{LE - 25}{85 - 25}$	Higher LEI, Lower vulnerability	http://www.dams.org/docs/kbase/studies /drafts/thscope.pdf
17	Er	Economic Recovery	FVI$_{Ec}$	R	#		The higher #, the high vulnerability	reffer to table
18	FI	Flood Insurance	FVI$_{Ec}$	R	-	the number flood insurances, if 0 than take 1	higher # of FI, lower vulnerability	
19	A$_{mInv}$	Amount of investment	FVI$_{Ec}$	R	-	Ratio of investment over the total GDP	Higher investment lower vulnerability	UNEP, 2004, IKONE project, Aktionsplan Hochwasser Neckar
20	D_L	Dikes , Levees	FVI$_{Ec}$	R	km/km	Km of dikes/levees over total length of river	Longer D_L, lower vulnerability	
21	D_Sc	Dams_Storage capacity	FVI$_{Ph}$	R	m	amount of storage capacity over area of sub-catchment	higher capacity, lower vulnerability	www.ucowr.siu.edu
22	R$_{ainfall}$	Rainfall	FVI$_{En}$	E	m/year	the average rainfall/year of a whole RB $\frac{mm}{1000 * year} = \frac{m}{year}$	Higher rainfall, higher vulnerability	http://www.dams.org/docs/kbase/studies /csthanx.pdf, Ekstrom et al.
23	D$_A$	Degrated Area	FVI$_{En}$	E	%	% of degraded area	Bigger D$_A$, higher vulnerability	WRI
24	U$_G$	Urban Growth	FVI$_{En}$	S	%	% of increase in urban area in last 10 years;	fast urban growth may result in poor quality housing and thus make people more vulnerable	UNDP/BCPR, 2004
25	L$_U$	Land Use	FVI$_{En}$	E	%	% of forested area	The higher %, the low vulnerability	http://www.mekongnet.org/images/5/5b/ Uruya.pdf, Tizsa River Basin Economic Development Programme
26	Ev	Evaporation rate	FVI$_{En}$	S	m/year	yearly evaporation rate	higher Ev, higher vulnerability	Mekong River Commission, Ekstrom et al.
27	N$_R$	Natural Reservation	FVI$_{En}$	S	%	% of natural reservation over total SC area $\frac{A_{ix}}{Total\_Area\_of\_River\_Basi} * 100$	Higher %, Lower vulnerability	UNEP, http://www.iatp.md/arii/TEXT/RO/internati onale/ariiRO.htm
28	U$_{npop}$	Unpopulated Area	FVI$_{En}$	E	%	% of area with density of population less than 10 pers/km^2	Higher Unpop area. Lower vulnerability	Water Resource Atlas
29	T	Topography	FVI$_{Ph}$	E	-	average slope of sub-catcment	The steeper slope, higher vulnerability	wikipedia, google earth, http://www.hydro.washington.edu
30	R$_D$	River Discharge	FVI$_{Ph}$	E	m^3/s	maximum discharge in record of the last 10 years, m^3/s	higher RD, higher vulnerability	http://www.dams.org/docs/kbase/studies /csthanx.pdf, Tisza River Action Plan, Authority of Water, Romania
31	Fo	Frequency of occurrence	FVI$_{Ph}$	E	years	years between floods	bigger 3 of years, high vulnerability	http://www.dams.org/docs/kbase/studies /csthanx.pdf, Tizsa River Action Plan, Authority of Water, Romania
32	Ev/R$_{ainfall}$	Evaporation rate/Rainfall	FVI$_{Ph}$	E	-	Yearly Evaporation over yearly rainfall	Higher the Ev, lower vulnerability	http://www.dams.org/docs/kbase/studies /csthanx.pdf, Ekstrom et al.
33	D_Sc	Dams_Storage capacity	FVI$_{Ph}$	R	m	amount of storage capacity over area of sub-catchment	higher m, higher vulnerability	http://www.dams.org/docs/kbase/studies /csthanx.pdf
34	AvRd	Average River Discharge	FVI$_{Ph}$	E	m3/s	average river discharge at the mouth		S. Djordjevic et al, IKONE project, Aktionsplan Hochwasser Neckar, www.dams.org
35	Sc/Vyear	Storage capacity over yearly discharge	FVI$_{Ec}$	R	m3/m3	Storage capacity divided by yearly volume runoff	higher Sc means lower vulnerability	Reffer to 35 & 36

Appendix 3.3. Urban Area Scale Indicators

No	Abb.	Name	Sub-index	FV Factor	Units	Definition of indicator	Functional relationship with vulnerability	Data Source
							Urban Scale	
1	PD	Population density	FVI_S	E	people/km2	There is an important exposure to a given hazard if population is concentrated	Higher # of people, higher vulnerability	http://www.dams.org/docs/kbase/studies/cst hanx.pdf
2	P_{FA}	Population in flood prone area	FVI_S	E	people	Number of people living in flood prone area	The higher number of people, higher vulnerability	http://www.dams.org/docs/kbase/studies/cst hanx.pdf
3	C_H	Cultural Heritage	FVI_S	E	-	number of historical buildings, museums, etc., in danger when flood occurs	high # of CH, higher the vulnerability	Municipalities
4	P_G	Population growth	FVI_S	E	%	% of growth of population in urban areas in the last 10 years	fast PG, higher vulnerability, hypothesis is made that fast population growth may create pressing on housing capacities	Municipalities
5	% of disable	Disable People	FVI_S	E	%	% of population with any kind of disabilities, also people less 12 and more than 65	higher %, higher vulnerability	INTUTE
6	HDI	Human Development Index	$FVI_{S,E}$	S	-	$HDI = \frac{1}{3}(LEI) + \frac{1}{3}(EI) + \frac{1}{3}(GI)$	The higher value, lower vulnerability	UNDP, 2004
7	C_M	Child Mortality	FVI_S	S	-	Number of children less than 1 year old, died per 1000 births	The higher number of chilren, higher vulnerability	http://www.dams.org/docs/kbase/studies/cst hanx.pdf, EPI
8	P_E	Past Experience	FVI_S	R	people	# of people who have been affected in last 10 years because flood events;	The higher value, lower vulnerability	http://68.178.233.117/MPcomp/2003/maps/A cademicindividual-Caquard.pdf, Selected
9	A	Awareness&Prepared ness	FVI_S	R	-	Range between 1-10	10 means lower vulnerability	reffer to table, appendix 4
10	D	Drainage	FVI_S	R	-	Range between 1-10	10 means lower vulnerability	Municipalities
11	C_{PR}	Communication Penetration Rate	FVI_S	R	%	% of households with sources of information	Higher percentage means lower vulnerability	INTUTE
12	S	Shelters	FVI_S	R	#/km²	number of shelters per km² including hospitals	bigger # of S, lower vulnerability	Municipalities
13	W_S	Warning system	FVI_S	R	-	if No W_S than the value is 1, if yes than the value is 10	Having WS reduces the vulnerability	Y/N
14	E_S	Emergency Service	FVI_S	R	#	number of people working in this service	bigger # of people, less vulnerable they are	Municipalities
15	E_R	Evacuation Roads	FVI_S	R	%	% of asphalted roads.	The better the quality of roads, improves the evacuation during floods	INTUTE
16	Ind	Industries	FVI_E	E	#	# of industries or any types of economic activities in urban area	The higher %, the high vulnerability	Municipalities
17	P_R	Proximity to river	FVI_E	E	km	average proximity of populated areas to flood prone areas	close to the river, higher vulnerability	google Earth
18	U_M	Unemployment	FVI_E	S	%	$U_M = \frac{\#of\_people Unempl}{Total\_Pop\_AptToWork} * 100$	The higher %, the high vulnerability	World Factbook 2005
19	I_{neq}	Inequality	FVI_E	S	-	Gini Coefficient for wealth inequality, between 0 and 1	Where 1 means low vulnerability	UN
20	FI	Flood Insurance	FVI_E	R	-	the number flood insurances, if 0 than take 1	higher # of FI, lower vulnerability	Insurance Companies
21	A_{mInv}	Amount of Investment	FVI_E	R	-	Ratio of investment over the total GDP	Higher the investment lower vulnerability	UNEP, 2004, IKONE project, Aktionsplan Hochwasser Neckar
22	D_L	Dikes_ Levees	FVI_E	R	km	Km of dikes/levees	Longer D_L, lower vulnerability	Water Authorities
23	D_Sc	Dams_Storage capacity	FVI_E	R	m³	Storage capacity in m3 of dams, polders, etc., upsteam of the city	higher m3, higher vulnerability	http://www.dams.org/docs/kbase/studies/cst hanx.pdf, Ekstrom et al.
24	R_D	River Discharge	FVI_E	E	m³/s	recorded river discharge at past flood events, m³/s	higher RD, higher vulnerability	Water Authorities
25	RT	Recovery time	FVI_E	R	days	Amount of time needed by the city to recover to a functional operation after flood events	the higher amount of time, the higher vulnerability	See Appendix 5
26	$R_{ainfall}$	Rainfall	FVI_{En}	E	m/year	the average rainfall/year $R_{ainfall} = \frac{mm}{1000 * year} = \frac{m}{year}$	Higher rainfall, higher vulnerability	Ekstrom et al.
27	L_U	Land Use	FVI_{En}	E	%	area destined for green areas inside the urban area	The higher %, the low vulnerability	http://www.mekongnet.org/images/5/5b/Uruy a.pdf, Tizsa River Basin Economic Development Programme
28	UG	Urban Growth	FVI_{En}	S	%	% of increase in urban area in last 10 years	fast urban growth may result in poor quality housing and thus make people more vulnerable	Municipalities
29	G_{WL}	Groundwater Level	FVI_{En}	S	m/year	yearly decrease rate in groundwater level	higher GWL, higher vulnerability	Water Authorities
30	T	Topography	FVI_{Ph}	E	-	average slope of sub-catcment	The steeper slope, higher vulnerability	wikipedia, google earth, http://www.hydro.washington.edu
31	R_D	River Discharge	FVI_{Ph}	E	m³/s	maximum river discharge in recor of the last 10 years, m³/s	higher RD, higher vulnerability	Water Authorities
32	RP	Return Periods	FVI_{Ph}	E	years	years between floods	bigger 3 of years, high vulnerability	Water Authorities
33	$E_V/R_{ainfall}$	Evaporation rate/Rainfall	FVI_{Ph}	E	-	Yearly Evaporation over yearly rainfall	Higher the Ev, lower vulnerability	Ekstrom et al.
34	D_Sc	Dams_Storage capacity	FVI_{Ph}	R	m³	amount of storage capacity	higher m, higher vulnerability	http://www.dams.org/docs/kbase/studies/cst hanx.pdf, Ekstrom et al.

CHAPTER 4

Reducing the complexity of Flood Vulnerability Index

This chapter has been published as:

Balica, S.F., Wright, N.G., 2010, Reducing the complexity of Flood Vulnerability Index, Environmental Hazards Journal, Volume 9, Number 4, 2010, pp. 321-339, Publisher Earthscan

4.1. Introduction

As seen in Chapter 3, the current flood vulnerability index (FVI) methodology uses 71 indicators in its calculation. However, it is recognized that some of these indicators may be redundant or have no influence on the results. This chapter presents the results of analysis carried out to select the most significant indicators in order to establish parsimonious usage of the FVI.

Like the original methodology, this is applicable at three different spatial scales (river basin, sub-catchment and urban) and to the various components of flood vulnerability (social, economic, environmental and physical). For the FVI methodology to be sustainable, improvements were made by analysing the indicators' relevance and by studying the main indicators (Section 4.3) needed to portray reality of the fluvial floods in an effective way. For this purpose, mathematical tools (a derivative and a correlation method) and expert knowledge (via a questionnaire) were used (Section 4.4).

Finally, all these methods were combined in order to select the most significant indicators and to simplify the FVI equations (Section 4.5). After reducing its complexity, the FVI can be more easily used as a tool for education, improvement of decision making and ultimately reduction of flood risk.

4.2. Background

This index follows on from other index-based approaches described below. Such an approach can assist in decision making in flood risk management by allowing decision makers to gain an insight in the following ways:

- The FVI can be used as a tool to communicate a multidisciplinary topic in a relatively straightforward way. Due to the large number of components, it gives a good overview of flood vulnerability at different scales. It also provides the user with a value that can be communicated to other stakeholders in a relatively simple way, and therefore should raise awareness for the topic of vulnerability.
- The FVI can inform decision makers and the general public about climate change risks, in order to increase their capacity to implement any necessary adaptation measures and to strengthen the resilience of a particular community.
- The holistic view provided by the FVI helps decision makers to determine and explore the possible harm and recognize the correct actions that can be taken before this harm occurs. It also helps when considering long-term flood management and shows how different factors may influence vulnerability.
- Vulnerability assessment is designed to produce certain information for specific target areas and constitutes part of the early warning system at all levels and components.
- The FVI may be an important tool in the context of strategic impact analysis to communicate impacts and vulnerability and to evaluate development alternatives for adaptation to the changes.
- An FVI distribution map can be generated and used as a measure for prioritizing adaptation.

The FVI methodology used in this thesis assesses the level of vulnerability for each factor (exposure, susceptibility and resilience) and of different spatial scales, and gives a quantitative evaluation by aggregating indicators. While exposure, susceptibility and resilience are considered as factors influencing vulnerability, and consequently the FVI, vulnerability itself is considered as having four components that relate to different ways in which the system is vulnerable. The interaction between the vulnerability factors and the vulnerability components serves as the foundation of FVI methodology, as explained in Section 3.2.1.1.

To remind, in this thesis, the vulnerability of a system to flood events is expressed by the following general equation:

$$\text{Vulnerability} = \text{Exposure} + \text{Susceptibility} - \text{Resilience} \qquad 4.1$$

The methodology presented here has been applied to case studies at various spatial scales (Balica et al., 2009a), which resulted in interesting observations on how vulnerability can be reflected by quantifiable indicators. In order to simplify the methodology and make it more readily understood, this chapter centers on reducing the number of indicators used for the FVI through different mathematical and social techniques, that is, sensitivity, correlation and expert survey. Firstly, the chapter focuses on FVI indicators and some of the current vulnerability indices; secondly, the chapter focuses on the analysis of FVI indicators and the use of different methods to find the most significant indicators; thirdly, all these methods are combined in order to simplify the FVI methodology, and the equations of each spatial scale and the remaining FVI indicators are presented; and finally, some discussion, conclusions and implications are given.

4.3. Reducing the number of FVI indicators

About 80 possible indicators were examined for upgrading the FVI (Balica et al, 2009a), taking into account the geographical scales: river basin, sub-catchment and urban. Forty indicators were included in the FVI computation; the rest were taken out of the equations due to redundancy of definitions, low relevance in flood vulnerability or difficulty in obtaining the required data.

These 40 indicators were then studied to portray reality in an effective way and to create a more discernible methodology that is perceived as realistic by stakeholders and the public.
The modified FVI methodology (previous chapter) shows that vulnerability includes three factors: exposure, susceptibility and resilience. The first two factors derive from social sciences, and relate to why a (socio-economic) system responds in a given way. The third factor, resilience, describes how a system reacts to a disturbance (ASCE & UNESCO, 1998). The mathematical theory of the index has to tie in harmoniously with the actual situation, to achieve a more objective FVI.

For river basin scale, 27 indicators were selected, which cover different dimensions of flood vulnerability and give an overview of the situation.

However, there are some limitations: on the one hand, having so many indicators makes it difficult to integrate the 'dynamics' in a model in that the indicators simplify the complexity of the actual situation, and on the other hand, at a river basin scale, the set of indicators chosen seems to cover quite well the social dimension of FVI (population in flood-prone area (PFA), human development index (HDI), child mortality (CM), past experience (PE), awareness and preparedness (A), communication penetration rate (CPR), warning system (WS), unemployment (U), evacuation roads (ER); see Appendix 4.1 for full details). The indicator PFA is relevant to the index, because the more the people living in a flood-prone area, the more exposure there is. The indicator PE reflects how a population with previous experience in dealing with floods acquires, accumulates and transfers knowledge on how to deal with flooding and its consequences. The indicator CPR is relevant because it facilitates awareness raising and preparedness and also facilitates the dissemination of information about responding to a flood. The CPR indicator is connected to the warning system indicator WS, and the more advanced and early the warning is, the more the population is able to respond and the further the consequences are reduced. Knowing about these warnings, the population can evacuate using the evacuation roads indicated in evacuation plans.

The environmental and physical set of indicators gives a good overview of soil and water conditions as well as land coverage. Two further scales are examined here: the sub-catchment scale (37 indicators) and the urban area scale (34 indicators). The choice of indicators used to assess vulnerability is the same for both spatial scales in about 26 instances, reflecting their similarity. At these scales, more variables need to be used or can be used to determine vulnerability more precisely. In the case of the sub-catchments, most of the indicators help to describe the rural living situation with respect to the risk of flooding.

For example, the following indicators are used to assess vulnerability at the sub-catchment scale but are not used on the urban scale: urbanized area, rural population, proximity to river, life expectancy, degraded area, land use (forest cover), natural reservation, unpopulated area and frequency of occurrence. On the other hand, the following indicators are used to assess vulnerability at the urban scale, but are not used at the sub-catchment scale: cultural heritage, population growth, shelters, emergency services, industries, contact with river, recovery time and drainage system.
The occurrence of a flood event at the lower spatial scales exposes proportionally more people to the effects of the flood. More people would have to be moved to shelters, more emergency services would be needed and more economic assets would be exposed to damage.

In general, the indicators for flood vulnerability at the lower scales need to reflect the resilience of the area, that is, the ability of society to cope or adapt during or after a flood event. Hence, it would be necessary for indicators to reflect whether shelters and emergency services are readily available, and also the proximity of the city to the river, the adequacy of the drainage system, the recovery time and exposure of the city's economic assets to floods. These indicators may not be a priority or that relevant on the sub-catchment scale. They can be extremely important, however, since FVI is aimed at assisting with decision making; thus recognizing the role of floodplains can lead to setting planning limitations to decrease vulnerability.

In the case of urban areas, other factors play a role as well, like cultural heritage or industries. Inundation can cause economic damage if important buildings or industries are affected. Overall, it can be seen that the indicators discussed above provide valuable information regarding the identification of the potential for flood vulnerability. Knowledge about each scale and the judgement of the planner should be used to make the correct decisions. It is important to note that a single indicator cannot assess the FVI or its potential for development. For a complete picture of any given situation, several indicators such as the ones presented here need to be quantified.

4.4. Identifying the most significant indicators

Due to its complexity, the methodology needs to reduce the number of indicators used because the ultimate aim is to provide the stakeholders with a clear and flexible methodology to evaluate flood vulnerability, in order to be used at various scales and in as many case studies as possible.

Many communities start prevention strategies without a real knowledge of the territory's vulnerability (Moris-Oswald and Sinclair, 2005; Barroca et al., 2006). The FVI will be useful if it is based on an indicator methodology and not only a subject of discussion after flooding events. For this reason, the focus after obtaining the FVI should be on the need for participation of all stakeholders in flood vulnerability discussions.

Three methods are presented below to assist in finding the most significant indicators belonging to the three different spatial scales and implicitly to the four components. This allows end users to simplify their work as much as possible while demonstrating their implementation in relevant case studies.

4.4.1. Differentiation method and discussion

The gradient of a scalar field is a vector field that points in the direction of the greatest rate of increase of the scalar field, and whose magnitude is the greatest rate of change.
The gradient of a scalar function f(x) with respect to a vector variable $x = \{x_1, . . ., x_n\}$ is denoted by ∇f, where ∇ (the nabla symbol) denotes the vector differential operator, del. The gradient of f is defined to be the vector field whose components are the partial derivatives of F:

$$\nabla f = \left(\frac{\partial f}{\partial x_1}, ..., \frac{\partial f}{\partial x_n} \right) \qquad 4.2$$

In order to use this methodology, all indicator values were normalized using predefined minimum and maximum values for each indicator. After computing the indicator derivatives, for easier interpretation the derivative values were also normalized, (Eq. 3.2, Chapter 3)

The derivative method was used for all three spatial scales and for the four components of vulnerability. Out of 12 instances (3 spatial scales × 4vulnerability components), for the sake of brevity, the case of river basin scale and social component is given here. In order to see which projections of the gradient are more relevant, see Table 4.1, where the normalized values are shown (the Euclidean norm was used; the Euclidean norm of a matrix A is EN = $\sqrt{\sum_{ij} \sqrt{a_{ij}^2}}$.

where a_{ij} is the ith row and the jth column of matrix A).

$$FVI_{social} = \left[\frac{P_{FA} * C_M * U_M}{P_E * AP * C_{PR} * HDI * W_S * E_R} \right] \qquad 4.3$$

$$\nabla FVI_{social} = \begin{bmatrix} \dfrac{C_M * U_M}{Pe * AP * CPR * HDI * Ws * E_R} \\[2mm] \dfrac{P_{FA} * U_M}{Pe * AP * CPR * HDI * Ws * E_R} \\[2mm] \dfrac{P_{FA} * C_M}{Pe * AP * CPR * HDI * Ws * E_R} \\[2mm] -\dfrac{P_{FA} * U_M * C_M}{Pe^2 * AP * CPR * HDI * Ws * E_R} \\[2mm] -\dfrac{P_{FA} * U_M * C_M}{Pe * AP^2 * CPR * HDI * Ws * E_R} \\[2mm] -\dfrac{P_{FA} * U_M * C_M}{Pe * AP * CPR^2 * HDI * Ws * E_R} \\[2mm] -\dfrac{P_{FA} * U_M * C_M}{Pe * AP * CPR * HDI^2 * Ws * E_R} \\[2mm] -\dfrac{P_{FA} * U_M * C_M}{Pe * AP * CPR * HDI * Ws^2 * E_R} \\[2mm] -\dfrac{P_{FA} * U_M * C_M}{Pe * AP * CPR * HDI * Ws * E_R^2} \end{bmatrix} \qquad 4.4$$

Please see Appendix 4.1 for details.

Using this approach, the FVI $_{social}$ river basin results are as follows.

In this analysis the magnitude of the gradient is taken to indicate the significance of the indicator.

Table 4.1. Derivative indicators of FVI $_{social}$ river basin

Derivative after	Raw Values			Relative Values*				
	Rhine	Danube	Mekong	Rhine	Danube	Mekong	Average	
PFA	0.119	0.118	0.279	0.004	0.007	0.064	0.025	The biggest slope, the most significant indicator
CM	**29.175**	**16.283**	**4.387**	**1.000**	**1.000**	**1.000**	**1.000**	
UM	0.096	0.142	1.032	0.003	0.009	0.235	0.082	
PE	**4.365**	**1.285**	**0.545**	**0.150**	**0.079**	**0.124**	**0.118**	
AP	0.059	0.073	0.136	0.002	0.005	0.031	0.013	
CPR	0.035	0.084	0.596	0.001	0.005	0.136	0.047	
HDI	0.032	0.055	0.163	0.001	0.003	0.037	0.014	
WS	0.030	0.051	0.109	0.001	0.003	0.025	0.010	
ER	0.030	0.085	0.227	0.001	0.005	0.052	0.019	
Complete	0.030	0.051	0.109	0.001	0.003	0.025	0.010	

·Normalized values computed with equation (3.2).

For example, for the social component of the river basin, the indicators child mortality and past experience are not part of the trend (see Figure 4.1). If we remove the two indicators from Figure 2, the Rhine river basin is the most socially vulnerable and Mekong is the least vulnerable from the three case studies analysed, which is not the case.

In reality, socially after computing FVI, the Rhine river basin has less exposure, less susceptibility and high resilience compared with the Mekong river basin. Therefore, the indicators child mortality and past experience are the most significant by this measure.

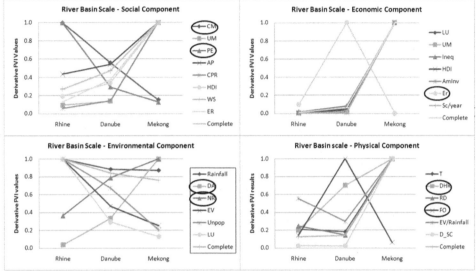

Figure 4.1. River basin scale – results of the FVI gradient indicators

Considering each of the vulnerability factors, the Mekong river basin is the most exposed (large numbers of people are living in the flood-prone area), most susceptible (unemployment, child mortality and HDI are high compared with the Rhine river basin) and least resilient (the kilometers of evacuation roads are very little and the communication penetration rate is low), even though in some factors the difference is not large, as is the case with the susceptibility of all river basins. The high resilience of the Rhine river basin is mainly due to communication penetration rate and evacuation roads, which are more than double the values for the Mekong river basin.

In summary, the most significant indicators at river basin scale are the following: for the social component child mortality (susceptibility) and past experience (resilience), for the economic component economic recovery (resilience), for the environmental component degraded areas and natural reservations, and for the physical component days with heavy rainfall (exposure) and flood occurrence (susceptibility); see Figure 4.1.

While the derivative method can be seen to have clearly identified significant indicators, it is a purely mathematical technique and hence should be considered alongside other methods as described below.

4.4.2. Survey method and discussion

A questionnaire was devised with four main questions, concerning the degree of significance of each indicator for each spatial scale and for the vulnerability components. Respondents were asked to rate the indicators on a scale from 5 to 1, where 5 indicates very high influence and 1 very low influence.

The questionnaire can be accessed at www.unesco-ihe-fvi.org (Balica and Wright, 2009). It was completed by 72 samples, including students of the UNESCO-IHE Institute for Water Education, staff and alumni of the institute, ICHARM, the Japanese Water Agency, Munich Re, Joint Research Center (Italy), Unisfera (Canada), US Geological Survey and different River Basin Organisations (see Appendix 4.2).

Overall, this gave a wide range of individuals in the water-related field who are involved in the decision-making process in different projects. The respondents evaluated the indicators' influence on flood vulnerability in general rather than on the index, thereby allowing for selection of significant vulnerability indicators based on expert assessment of vulnerability.

The results of the questionnaire for the four components are enunciated below for the river basin scale.

Social component: more than 90 per cent of 64 people answered that population in a flood prone area (PFA) is very important and important; more than 60 per cent answered that rural population, child mortality, trans-boundary river commissions, past experience and awareness/ preparedness indicators are important and that the indicator unemployment is not so significant.

Economic component: out of 68 people 91 per cent selected the land-use indicator as having very high influence; amount of investment, economic recovery, quality of infrastructure and human development index were all rated as having high influence. However, the indicator inequality was rated as low influence.

Environmental component: for the rainfall indicator 72 people (100 per cent) answered high influence (64 per cent said very high influence). The indicator degraded area is also considered important for the environment component by 67 per cent, the same as for forested area and type of vegetation. Forty-seven per cent felt natural reservations was important and 35 per cent evaporation. The indicator unpopulated area was assessed as having low significance by the expert community. This actually makes sense as this indicator is not relevant for the environmental component: the population in a flood-prone area PFA is an inversely proportional indicator, which is highly relevant for the social component. Therefore, we decided to remove this indicator for the environmental component and to analyse the impact of this change on the results (see later).

Physical component: out of 68 persons 94 per cent said that topography has very high and high influence; 90 per cent said that river discharge highly influences vulnerability. More than 70 per cent answered that indicators heavy rainfall, flood occurrence, flood duration and dam storage capacity have high influence on vulnerability.

After analysing the answers and aggregating the results, it was noticed that in the users' opinion the following indicators are not considered significant for the FVI: social health of an economy (unemployment), practical and societal grounds (life expectancy index and inequality) and the unpopulated area, cadastre survey and evaporation rate (see Appendix 4.1 for definitions). Although the users' answers can be subjective, we chose to include these results based on the wide knowledge and experience of the respondents and with the intention of considering the insights alongside those from more quantitative analysis. Most of the questionnaire answers confirmed the derivative method results and also the judgments of the author.

4.4.3. Correlation method and discussion

Correlation is a common and useful statistical method and was preferred over principal component analysis (PCA) (Jolliffe, 2002) in this work because of insufficient case studies for objective results with PCA. A correlation coefficient is a single number that describes the degree of relationship between two variables. Suppose we have two variables X and Y, with means X and Y, respectively, and standard deviations S_X and S_Y, respectively. The Pearson correlation is computed as:

$$r = \frac{\sum_{i=1}^{n}(X_i - \overline{X})(Y_i - \overline{Y})}{(n-1)S_X S_Y}$$

4.5

The correlation coefficient is always between -1 and +1. The closer the correlation is to +1, the closer the two variables to a perfect linear relationship. A strong positive association is interpreted as a correlation value between +0.7 and +1.0 (Simon, 2005).

In order to compute the Pearson correlation between the flood vulnerability indicators, for each spatial scale multiple case studies were used. For river basin scale, four more case studies were considered: Loire (Europe), Amazon (Latin America), Nile (Africa) and Limpopo (Africa) river basins. These new case studies were selected due to their representativeness and their diversity, with the goal of having more cases for better correlation results.

For the sub-catchment scale, 18 more case studies were taken into account for all the sub catchments of the Philippines archipelago. These sub-catchments were selected in order to compare the FVI methodology with an earlier one (Connor and Hiroki, 2005) applied also in the same region.

For the urban scale four more cases were chosen: Delft (the Netherlands), Drobeta Turnu Severin (Romania), Tours (France) and Dordrecht (the Netherlands). As with the river basin case studies, these cities were selected for their diversity.

The complete FVI results of all case studies presented here can be seen online at http://unesco-ihe-fvi.org. The key features pertinent to this paper are given below.

For the river basin example, the results of the strongest correlations between the indicators at all vulnerability components can be seen below. Strong positive associations can be observed between the following indicators:

- social component: communication penetration rate and evacuation roads r = 0.92, unemployment and child mortality r +0.75, and past experience and child mortality r = 0.70;
- economic component: land use and economic recovery r = 0.72, inequality and amount of investment r = 0.75, storage capacity and economic recovery r = 0.78, and storage capacity and amount of investments r = 0.81;
- environmental component land use and unpopulated area r = 0.84, natural reservation and rainfall r = 0.70, natural reservation and unpopulated area r = 0.83, and land use and natural reservation r = 0.84;
- physical component: days with heavy rainfall and flood occurrence r = 0.99, flood occurrence and dam storage capacity r = 0.83, and days with heavy rainfall and dam storage capacity r = 0.78.

In view of the significant correlation between some indicators, these strongly correlated indicators are not independent and by choosing only one it is clear that in each group we are able to reduce the total number without affecting the FVI.

For the river basin scale, the following indicators were removed:

- social component: communication penetration rate, correlated to evacuation roads; unemployment, correlated to child mortality;
- economic component: land use, correlated to unpopulated area; storage capacity, correlated to economic recovery;
- environmental component: unpopulated area, correlated to natural reservation;
- physical component: flood occurrence, correlated to the dam storage capacity.

4.5. Combining results for a simplified FVI methodology

The selection of the most significant indicators was done by using the following selection process (see Figure 4.2):

- the most significant indicators after using the derivative method were selected;
- the most significant ones after using the questionnaire were selected;
- after determining the correlated indicator pairs, only the most significant ones were kept;
- finally, indicators not selected from the above three methods were removed.

Figure 4.2. The method used to reduce the number of FVI indicators

4.5.1. River basin scale

The updated equations for all components of the river basin scale after the selection process are:

$$FVI_{social} = f\left[\frac{P_{FA} * C_M * U_M}{P_E * AP * C_{PR} * HDI * W_S * E_R}\right]; \quad updatedFVI_{social} = f\left[\frac{P_{FA} * C_M}{P_E * AP * W_S * E_R}\right] \qquad 4.6$$

$$FVI_{economic} = f\left[\frac{L_U * U_M * I_{neq} * HDI}{AmInv * E_R * Sc / yeardisch\arg e}\right]; \quad updatedFVI_{economic} = f\left[\frac{HDI * Ineq}{AmInv * E_R}\right] \qquad 4.7$$

$$FVI_{environmental} = f\left[\frac{R_{a\inf all} * D_A}{N_R * E_V * U_{npop} * L_U}\right] \quad updatedFVI_{environmental} = \left[\frac{R_{a\inf all} * D_A}{N_R * E_V * L_U}\right] \qquad 4.8$$

$$FVI_{physical} = f\left[\frac{T * D_{HR} * R_D * F_o}{E_V / R_{a\inf all} * D\_S_C}\right]; \quad updatedFVI_{physical} = f\left[\frac{T * D_{HR} * R_D * F_o}{D\_S_C}\right] \qquad 4.9$$

For definitions of the indicators the reader is referred to Appendix 4.1.

The total FVI is a composite additive index based on the four vulnerability components.

Figure 4.3 show that three components (social, economic and physical) maintain their form after updating the methodology. However, the environmental component changes with the update: the Mekong basin becomes the most environmentally vulnerable to flooding.

The only difference between the two equations for the environmental component is the unpopulated area indicator. In the derivative method analysis this indicator was not highlighted as significant, while in the questionnaire it had a low score. In the correlation method this indicator was correlated with other indicators. In view of this it was removed, but it does still seem to be influential in some circumstances.

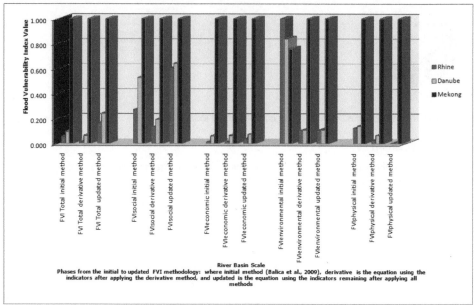

Figure 4.3. Comparison between the different phases of FVI equations – river basin scale

4.5.2.. Sub-catchment scale

The results on the sub-catchment scale are:

$$FVI_{social}=f\left[\frac{P_{FA}*R_{Pop}*\%_{disable}*C_M}{P_E*A/P*C_{PR}*W_S*E_R*HDI}\right]; \qquad updatedFVI_{social}=f\left[\frac{P_{FA}*R_{Pop}*\%_{disable}*C_M}{P_E*C_{PR}*W_S*E_R}\right] \qquad 4.10$$

$$FVI_{economic}=f\left[\frac{L_U*U_M*I_{neq}*U_A}{L_{EI}*F_I*AmInv*S_C/_{Vyear}*E_{CR}}\right]; \qquad updatedFVI_{economic}=f\left[\frac{L_U*U_A}{F_I*AmInv*S_C/_{Vyear}}\right] \qquad 4.11$$

$$FVI_{environmental}=f\left[\frac{R_{a\,inf\,all}*D_A*U_G}{L_U*E_V*N_R*U_{npop}}\right] \qquad updatedFVI_{environmental}=f\left[\frac{R_{a\,inf\,all}*D_A*U_G}{L_U*U_{npop}}\right] \qquad 4.12$$

$$FVI_{physical}=f\left[\frac{T}{E_V/_{R_{a\,inf\,all}}*S_C/_{Vyear}*D\_L}\right]; \qquad updatedFVI_{physical}=f\left[\frac{T}{E_V/_{R_{a\,inf\,all}}*S_C/_{Vyear}*D\_L}\right] \qquad 4.13$$

Again the reader is referred to Appendix 4.1.

The comparison between the existing FVI methodology and the updated one can be seen in Figure 4.3; the shape of the social, economic and physical component is maintained as in the existing FVI results, but the environmental component again varies.

As seen in Figure 4.4, the most environmentally vulnerable sub-catchment is the Tisza River; its nearest sub-catchment in the updated equation is the Bega River; the two sub-catchments and their rivers are very alike environmentally. The Neckar River has experienced environmental problems, due to large industries and lack of attention to environmental problems for many years. This way of thinking has changed in the last 20 years to a more environmentally friendly approach to river management; these improvements have contributed to the reduction of flood damage.

The main advantages of an FVI are the intellectual exercise in defining precisely what constitutes vulnerability for a particular region, and the possible ways of increasing resilience. Used by local stakeholders, FVIen can permit discussion on what is important and how one can reduce environmental flood vulnerability, particularly in the pro-active and prevention steps.

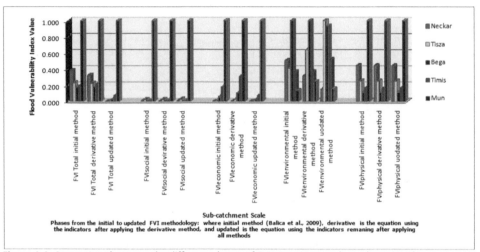

Figure 4.4. Comparison between the different phases of FVI equations – sub-catchment scale

4.5.3. Urban area scale

Urban areas are densely populated, which makes them especially vulnerable to flood effects. Our cities were established by early settlers along river banks. The derivative and correlation methods show that child mortality, past experience and cultural heritage are very significant.

The questionnaire shows that the population in flood-prone areas is most significant (97 per cent), with awareness and preparedness (90 per cent), warning systems and emergency services, evacuation roads (0.85 per cent) all being rated as influential. However, the cadastre survey indicator is not considered important.

With regard to the urban area example, are listed below the results of the strongest correlations between the indicators for all vulnerability components. Strong positive associations are observed between the following indicators:

- social component (population growth and shelters r = 0.85, population in flood-prone areas and child mortality r = 0.98, past experience and population in flood-prone areas r = 0.98, human development index and communication penetration rate r = 0.80, child mortality and past experience r = 0.99, awareness and warning system r = 0.81, communication penetration rate and evacuation roads r = 0.96, population growth and emergency service r = 0.75, human development index and evacuation roads r = 0.72, and human development index and shelters r = 0.71);

- economic component (industries and recovery time r = 0.84, dikes (levees) and amount of investment r = 0.87, dam storage capacity and recovery time r = 0.81, industries and inequality r = 0.73, unemployment and flood insurance r = 0.73, and inequality and recovery time r = 0.72);
- environmental component (no strong correlations);
- physical component (no correlation between the indicators).

$$FVI_{social}=f\left[\frac{P_D*P_{FA}*C_H*P_G*\%_{disables}*HDI*C_M}{P_E*AI P*C_{PR}*S*W_S*E_R*E_S}\right]; \quad updatedFVI_{social}=f\left[\frac{P_D*C_H*P_G*\%_{disables}*C_M}{P_E*AI P*S*W_S*E_R*E_S}\right] \quad 4.14$$

$$FVI_{economic}=f\left[\frac{I_{ND}*C_R*U_M*I_{neq}*U_G*HDI*R_D}{F_I*AmInv*D\_S_C*D*RT}\right]; \quad updatedFVI_{economic}=f\left[\frac{I_{ND}*C_R*U_M*U_G*HDI*R_D}{F_I*AmInv*D\_S_C*D}\right] \quad 4.15$$

$$FVI_{environmental}=f\left[\frac{U_G*R_{a\,inf\,all}}{E_V*L_U}\right] \qquad updatedFVI_{environmental}=f\left[\frac{U_G*R_{a\,inf\,all}}{E_V*L_U}\right] \qquad 4.16$$

$$FVI_{physical}=f\left[\frac{T*C_R}{E_V\Big/R_{a\,inf\,all}*S_C\Big/Vyear*D-L}\right]; \quad updatedFVI_{environmental}=f\left[\frac{T*C_R}{E_V\Big/R_{a\,inf\,all}*S_C\Big/Vyear*D-L}\right] \quad 4.17$$

Again the reader is referred to Appendix 4.1.

The comparison between the existing FVI methodology and the updated one for the urban area scale can be seen in Figure 4.5; the shape of the social, economic, environmental and physical components is maintained as in the initial FVI results, but the total FVI results vary, because the values of some components are changing. For example, the social component values are higher using the updated methodology compared with the initial one.

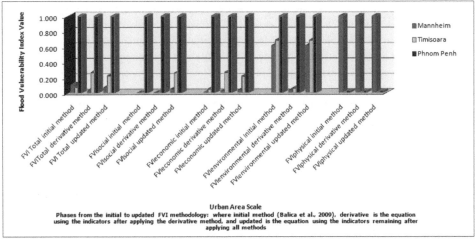

Urban Area Scale
Phases from the initial to updated FVI methodology: where initial method (Balica et al., 2009), derivative is the equation using the indicators after applying the derivative method, and updated is the equation using the indicators remaining after applying all methods

Figure 4.5. Comparison between the different phases of FVI equations – urban area scale

4.6. Conclusions

In this chapter, the overall number of indicators was reduced from the 71 indicators used initially to only 28 indicators that were retained: 20 for river basin scale, 22 for sub-catchment and 27 for urban area. A further reduction in indicators number will be taken into account in future work (e.g. coastal cities flood vulnerability index) so as to simplify the decision-making process and the effort of collecting the data. This important reduction was obtained after applying the derivative method, the correlation method and the questionnaire on the three spatial scales. By analysing the existing indicators for each spatial scale and for each vulnerability component, only the most significant ones were retained. Appropriate mathematical methods were applied (a derivative and correlation method) to sort out the essential key indicators for a simpler, easier and low-cost application. Analysing the indicator's significance through a survey was also carried out to portray reality in an effective way. This was done using a questionnaire with the participation of subjects from different water backgrounds and cultures (ICHARM, US Geological Survey, Japan Water Agency, UNESCO-IHE, River Basin Organisations, Joint Research Centre and diverse scientists). Those results were combined with the mathematical ones to obtain appropriate FVI indicators in order to easily formulate the FVI equations.

After combining these methods, it was noticeable that the environmental component of the river basin and sub-catchment is more rational as the flood vulnerability ranking changes for the case studies using the new equation. Once updated, the environmental set of indicators gives a better overview of the environmental FVI on water conditions as well as land coverage.

Using the updated equations will direct FVI users to a simplified usage and simpler understanding of the methodology, which can be used as a tool for decision making to direct investments to the most appropriate sectors and also to help in the decision-making process relating to flood defence, policies, measures and activities. The FVI approach allows, irrespective of uncertainties, relative comparisons to be made between spatial scales. While a level of uncertainty is inherent in FVI, the use of it in operational flood management is highly relevant for policy and decision makers in terms of starting adaptation plans. It offers a more transparent means of making such priorities, which inevitably are considered as highly political decisions. It may also be considered as a means to steer flood management policy in a more sustainable direction. However, as individual information is lost in the aggregation process, it needs to be retrieved by a more in-depth analysis of each process in order to design policies and their implementation.

Implementation includes options for preventing, reacting to, recovering or adapting to the impact of floods; hence it is useful to have an easy-to-apply instrument that can help give an overview of the main points by having one single and comparable number, the FVI. The FVI is necessary, but not sufficient, for decision making and therefore should be used in combination with other decision-making tools. This should specifically include participatory methods with the population of areas identified as vulnerable and should also include a team of multidisciplinary thematic specialists and knowledgeable societal representatives and those with expert judgment.

The updated methodology used on the three spatial scales has been disseminated via the FVI website (http://unesco-ihe-fvi.org), and the results serve as references to enable users to apply it in different parts of the world. Through this website, the FVI can help to assess and improve the links of the safety chain in risk management.

Appendices

Appendix 4.1. FVI indicators and their abbreviations

	Abb.	Indicators Name	Belonging to:	Definition of Indicators
			FVI Indicators and their abbreviations	
Social Component	PD	Population density	SC, UA	There is an important exposure to a given hazard if population is concentrated
	P$_{FA}$	Population in flood prone area	RB, SC, UA	Number of people living in flood prone area
	U$_A$	Urbanized Area	SC,	% of total area which is urbanized
	R$_{pop}$	Rural population	SC,	% of population living outside of urbanized area
	% of disable	Disabled People	SC, UA	% of population with any kind of disabilities, also people less 12 and more than 65
	CH	Cultural Heritage	UA	number of historical buildings, museums, etc., in danger when flood occurs
	HDI	Human Development Index	RB, SC, UA	$= \frac{1}{3}(LEI) + \frac{1}{3}(EI) + \frac{1}{3}(GI)$
	P$_G$	Population growth	UA	% of growth of population in urban areas in the last 10 years
	C$_M$	Child Mortality	RB, SC, UA	Number of children less than 1 year old, died per 1000 births
	P$_E$	Past Experience	RB, SC, UA	# of people who have been affected in last 10 years because flood events;
	S	Shelters	UA	number of shelters per km^2, including hospitals
	E$_S$	Emergency Service	UA	number of people working in this service
	A	Awareness&Preparedness	RB, SC, UA	Range between 1-10
	C$_{PR}$	Communication Penetration Rate	RB, SC, UA	% of households with sources of information
	W$_S$	Warning system	RB, SC, UA	if No W$_S$ than the value is 1, if yes W$_S$ than the value is 10
	E$_R$	Evacuation Roads	RB, SC, UA	% of asphalted roads.
Economic Component	LU	Land Use	RB, SC	% area used for industry, agriculture, any types of economic activities
	PR	Proximity to river	SC	average proximity of populated areas to flood prone areas
	Ind	Industries	UA	# of industries or any types of economic activities in urban area
	Cr	Contact with River	UA	Kms along the river bank
	UM	Unemployment	RB, SC, UA	$U_m = \frac{\#of\_people\_Unempl}{Total\_Pop\_AptToWork} * 100$
	Ineq	Inequality	RB, SC, UA	Gini Coefficient for wealth inequality, between 0 and 1
	LEI	Life expectancy Index	SC	$\frac{LE - 25}{85 - 25}$
	FI	Flood Insurance	SC, UA	the number flood insurances, if 0 than take 1
	AmInv	Amount of Investment	RB, SC, UA	Ratio of investment over the total GDP
	D_L	Dikes_ Levees	SC, UA	Km of dikes/levees
	D_Sc	Dams_Storage capacity	SC, UA	Storage capacity in m3 of dams, polders, etc., upsteam of the city
	RT	Recovery time	UA	Amount of time needed by the city to recover to a functional operation after flood events
	Er	Economic recovery	RB, SC	How affected is the economy of a region at a large time scale, because of floods
	Sc/Year	Storage capacity over yearly discharge	RB, SC	Storage capacity divided by yearly volume runoff
	U$_A$	Urbanized Area	SC	% of total area which is urbanized
Environmental Component	Rainfall	Rainfall	RB, SC, UA	the average rainfall/year of a whole RB $= \frac{mm}{1000 * year} = \frac{m}{year}$
	DA	Degrated Area	RB, SC	% of degraded area
	UG	Urban Growth	SC, UA	% of increase in urban area in last 10 years;
	LU	Land Use	RB, SC	% of forested area
	EV	Evaporation rate	RB, SC, UA	yearly evaporation rate $N_g = \frac{A_{mt}}{Total\_Area\_of\_River\_Basi} *100$
	NR	Natural Reservation	RB, SC	% of natural reservation over total RB
	Unpop	Unpopulated Area	SC,	% of area with density of population less than 10 pers/km^2
Physical Component	T	Topography	RB, SC, UA	average slope of river basin
	D$_{HR}$	# of days with heavy rainfall	RB,	number of days with heavy rainfall, more than 100mm/day
	RD	River Discharge	RB, SC, UA	maximum discharge in record of the last 10 years, m^3/s
	Fo	Frequency of occurance	RB, SC	years between floods
	Drain	Drainage system	UA	Kms of drainage pipes
	EV/Rainfall	Evaporation rate/Rainfall	RB, SC, UA	Yearly Evaporation over yearly rainfall
	Sc/Year	Storage over yearly runoff	UA	Storage capacity divided by yearly volume runoff
	D_Sc	Dams_Storage capacity	RB, SC, UA	The total volume of water, which can be stored by dams, polders, etc.
	AvRd	Average River Discharge	RB, SC, UA	average river discharge at the mouth

where RB is river basin scale, SC is sub-catchment scale and UA is urban area scale.

Appendix 4.2. Questionnaire groups

Questionnaire Groups	
Institute/Agency/Organisation	Number of respondents
UNESCO-IHE students/staff	56 out of 72
Other institutions	16 out of 72

Chapter 5

Parametric and physically based modeling techniques for flood risk and vulnerability assessment: a comparison

5.1. Introduction

As seen in Chapter 3 and Chapter 4 a flood vulnerability index (FVI) methodology was developed and applied. However, there is need to identify the risk of flooding in flood prone areas to support decisions for flood management from high level planning proposals to detailed design.

There are many methods available to undertake such studies, the most traditional, and commonly used, of which is computer modelling and inundation mapping and on the other hand the no collectively accepted parametric approach, the vulnerability assessment. The applicability of these approaches shows advantages and disadvantages for decision makers and how do the two approaches compare in use.

This chapter focuses on the applicability and performance of the flood vulnerability index methodology in comparison to the deterministic methods, see Section 5.4, methodology. This is undertaken in a data scarce area (Budalangi area, Kenya, using the SOBEK 1D/2D model, see section 5.4.2). The two data areas are compared in Section 5.6, where an analysis between the two approaches is done. This chapter comes to indicate that a combination of approaches should be used in flood risk planning and assessment in both data rich and data scarce contexts.

The chapter finishes with a discussion Section 5.6 and the conclusions Section 5.7. It is concluded that the parametric approach is not just something to be applied before or/and after the completion of the modelling work, but the FVI, through indicators is the only approach which evaluates vulnerability to floods; the deterministic approach has a better science based, but limited evaluation of vulnerability.

5.2. Background

Floods are one of the most common and widely distributed natural risks to life and property worldwide. Damage caused by floods on a global scale has been significant in recent decades (Jonkman and Vrijling, 2008). In 2011, floods were reported to be the third most occurring disaster, after earthquake and tsunami, with 5202 deaths, and affecting millions of people (CRED, 2012). Nature is a powerful force; river, coastal and flash floods can claim human lives, destroy properties, damage economies, make fertile land unusable and damage the environment. The development of techniques, measures and assessment methodologies to increase understanding of flood risk or vulnerability can assist decision makers greatly in reducing damage and fatalities. Different methods to assess risk and vulnerability of areas to flooding have been developed over the last few decades. This paper aims to investigate two of the more widely used methods. Traditional physically-based modelling approaches to risk assessment and parametric approaches for assessing flood vulnerability. Both methods have the overall purpose of discussing the benefits of each to decision makers.

5.2.1 Flood risk as a concept

The term "risk", in relation to flood hazards, was introduced by Knight in 1921, and is used in diverse different contexts and topics showing how adaptive any definition can be (Sayers et al., 2002). In the area of natural hazard studies, many definitions can be found. It is clear that the many definitions related to risk (Alexander, 1993; IPCC, 2001; Plate, E., 2002; Barredo et al., 2007) are interrelated and interchangeable and each of them has certain advantages in different applications (e.g. Sayers et al., 2002; Merz et al., 2007).

This study will consider risk as the product of two components, i.e. probability and consequence (Smith, 2004):

$$Risk = Probability \times Consequences$$ Eq. 5.1

This concept of flood risk is strictly related to the probability that a high flow event of a given magnitude occurs, which results in consequences which span environmental, economic and social losses caused by that event. The EU Flood Directive 2007/60/EC (EC, 2007) and UNEP, (2004) uses this definition of risk where "flood risk" means the combination of the probability of a flood event and of the potential adverse consequences for human health, the environment, cultural heritage and economic activity associated with a flood event

5.2.2. Hazard and Flood Hazard as a concept

"The probability of the occurrence of potentially damaging flood events is called flood hazard" (Schanze, 2006). Potentially damaging means that there are elements exposed to floods which may be harmed. Flood hazards include events with diverse characteristics, e.g. a structure located in the floodplain can be endangered by a 20-year flood and a water level of 0.5m and by 50-year flood and a water level of 1.2m. Heavy rainfall, coastal or fluvial waves, or storm surges represent the source of flood hazard. Generally these elements are characterised by the probability of flood event with a certain magnitude and other characteristics.

5.2.3. Vulnerability and Flood vulnerability as a concept

While the notion of vulnerability is frequently used within catastrophe research, researchers' notion of vulnerability has changed several times lately and consequently there have been several attempts to define and capture the meaning of the term. It is now commonly understood that "vulnerability is the root cause of disasters" (Lewis, 1999) and "vulnerability is the risk context" (Gabor and Griffith, 1980). Many authors discuss, define and add detail to this general definition. Some of them give a definition of vulnerability to certain hazards like climate change (IPCC, 2001), environmental hazards (Blaikie et al., 1994); (Klein and Nicholls, 1999), (ISDR, 2004), or the definition of vulnerability to floods (Veen & Logtmeijer 2005, Connor & Hiroki, 2005, UNDRO, 1982, McCarthy et al., 2001).

This study will use the following definition of vulnerability specifically related to flooding:
> The extent to which a system is susceptible to floods due to exposure, a perturbation, in conjunction with its ability (or inability) to cope, recover, or basically adapt.

5.3. The practice of flood risk and vulnerability assessment

Different methods to assess or determine hazard, risk and vulnerability to flooding have evolved through ongoing research and practice in recent decades. Two distinct method types can be distinguished and are considered in this paper:
- Deterministic modelling approaches which use physically based modelling approaches to estimate flood hazard/probability of particular event, coupled with damage assessment models which estimate economic consequence to provide an assessment of flood risk in an area.

- Parametric approaches which aim to use readily available data of information to build a picture of the vulnerability of an area.

Each method has developed from different schools of thought; the first approach mentioned is the traditional method which is routinely used in practice and academia alike. The second approach has evolved from several concerns such us: the internal characteristics of the system, global climate change and the political and institutional characteristics of the system. However, it takes a long time to develop the structural and non-structural measures required to prepare for flooding. In order to help guide such policy decisions, the development of a practical method for assessing flood vulnerability was needed. Among this need, this parametric approach points on vulnerability assessments to minimize the impacts of flooding and also to increase the resilience of the affected system.

5.3.1. The physically based modelling approach

Floods are primarily the result of extreme weather events. The magnitude of such an extreme event has an inverse relationship with the frequency of its occurrence i.e. floods with high magnitude occur less frequently than more moderate events. The relationship between the frequencies of occurrence and the magnitude of the extreme event is traditionally established by performing a frequency analysis of historical hydrological data using different probability distributions.

Once the frequency, magnitude and shape of the hydrograph are established, computer models which discretise the topographical river and land form are used to estimate flood depth, flood elevation and velocity (Hansson et al., 2008). Calculation of flood inundation depth and inundation extent is done using computational models based on solutions of the full or approximate forms of the shallow water equations. These types of models are one (1D) or two-dimensional (2D). 1D modelling is the common approach for simulating flow in a river channel, where water flow in the river is assumed to flow in one dominant direction which is aligned with the centre line of the main river channel. A 1D model can solve flood flows in open channels, if the shallow water assumptions that vertical acceleration is not significant and that water level in the channel cross-section is approximately horizontal are valid. However problems arise when the channel is embanked and water levels are different in the floodplain than in the channel and 2D models are needed in this situation. The hydraulic results from a computer model, such as inundation depth, velocity and extent can be used for loss estimation due to a particular design flood event. These parameters can then be linked to estimates of economic damage and loss in the affected area. Different models of damage and loss are available and are based on established economic relationships (ref).

This method relies on a significant amount of detailed topographic, hydrographic and economic information in the area studied. If the information is available, fairly accurate estimates of the potential risk to an area, as a result of economic losses, can be calculated. This type of flood hazard and associated economic loss information is reasonably easily communicated to the public. With the case of economic loss the public is used to hearing information provided in this manner. However, if the information for the model construction is not available, the method is likely to incur significant anomalies, which can call into question the validity of the assessment. These types of knowledge gaps and uncertainties are difficult to communicate effectively and can confuse decision makers and the public alike. The scientific community therefore has researched methods that will overcome these problems. In this context it becomes important to evaluate the hazard, risk and vulnerability to flooding also from a different perspective: the parametric approach.

5.3.2. The parametric approach

The parametric approach, introduced in 80's by Little and Robin, (1983), starts from the perspective of limited data, and has developed further since. The parametric approach aims to estimate the complete vulnerability value of a system by using only a few readily available parameters relating to that system, though the implementation of the approach is not simple.

Four types of parametric approaches have been developed by the scientific communities: i) estimating the complete vulnerability value of a system by using only few parameters relating to that system, ii) estimation of "the imputation of non-observable values" (Glynn et al., 1993), in which the observed parameters are used to model the non-observed ones. (This assumption can be wrong), iii) the "parametric modelisation via maximum likelihood" (Little and Rubin, 1987), which is not a direct approach and is based on large number of assumptions; and iv) the "semi-parametric approach" (Newey, 1990) which allows modelling only of what is strictly necessary.

This study considers the first type of parametric approach, where the indicators and results rely on assumptions that cannot be validated from the observed data. This parametric approach tries to design a methodology that would allow the experts to assess the vulnerability results depend on the system characteristics and also to show the drawbacks, the practical and the philosophical in the specifications of the likelihood function (Serrat and Gomez, 2001).

In a general context, vulnerability is constructed like an instrumental value or taxonomy, measuring and classifying social, economic and environmental systems, from low vulnerability to high vulnerability. The vulnerability notion has come from different disciplines, from economics and anthropology to psychology and engineering (Adger, 2006); the notion has been evolving giving strong justifications for differences in the extent of damage occurred from natural hazards.

Whatever the exact measure of vulnerability one chooses to work with, the starting point is to estimate the right parameters of the process under the specification of the datasets. Vulnerability assessments have to be explicitly forward-looking. No matter how rich the data, the vulnerability of various systems is never directly obvious.

At spatial and temporal scales, several methodologies such as parametric-based approaches are applied to a vast diversity of systems: Environmental Vulnerability Index (EVI), Pratt et al, 2004; The Composite Vulnerability Index for Small Island States (CVISIS), Briguglio, 2003; Global Risk and Vulnerability Index (GRVI), Peduzzi et al., 2001; Climate Vulnerability Index (CVI), Sullivan and Meigh, 2003, etc..

This chapter uses a parametric approach proposed by Balica et al., (2009) to determine and index flood vulnerability for four system components (social, economic environmental and physical).
The parametric approach has some drawbacks, such as: an inevitable level of assumptions, the need for a sensitivity analyses, reliable sources and the subjective manner of interpreting the results.

5.3.3. Comparison of approaches

Physically based modelling and parametric approaches offer two different techniques for assessing flood risk and vulnerability. In light of these two distinct approaches, a clear question arises: what are the different advantages and disadvantages for decision makers using these techniques and "how do the two approaches compare in use?"

In order to answer this research question it is important to assess what decision makers require from these techniques in order to reach decisions. For the purposes of this study the following key components are identified:
- Information on the mechanism and cause of flooding (flood hazard) in the area studied.
- Information on the health and safety implications for the affected population of the flood hazard posed in the area, and the relative areas or population who are particularly vulnerable (and why).
- Information on the economic damage and losses expected in the area given a particular event.

In addition to these key components a fourth criteria was identified:
- How easily is this information communicated, both
    - From the expert undertaking the study to the decision-maker and
    - From the decision-maker to the public

This study will use the above identified criteria to compare the application of the two techniques (physically based modelling and the parametric approach) to a case study area in Budalangi, on the Nzioa River in Western Kenya. The paper aims to investigate the benefits and drawbacks of each approach, with the purpose of informing decision makers of the use.

5.4. Methodology

The scope of the present chapter is to compare a parametric approach (Flood Vulnerability Index (FVI)) with traditional physically-based hydraulic modelling for flood risk analysis in order to determine the advantages of using one or the other in design and decision-making when flood hazard is involved. The general framework for the methodology is set out in Figure 5.1.

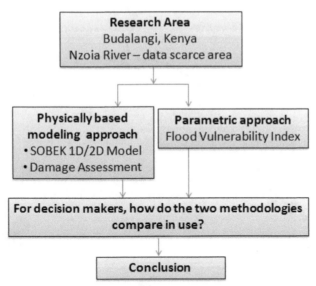

Figure 5.1. Proposed methodology

5.4.1 Case Study Area

The Nzoia River originates in the South Eastern part of Mt. Elgon and the Western slopes of Cherangani Hills at an elevation of about 2300 m.a.s.l and it is one of the major rivers flowing into Lake Victoria. Nzoia river basin covers an area of 12709 km^2 in Western Kenya (Figure 2). The Nzioa River discharges into Lake Victoria in Budalangi, Busia district. The river is of international importance, as it is one of the major rivers in Nile basin contributing to the shared water of Lake Victoria (NRBMI, (nd)).

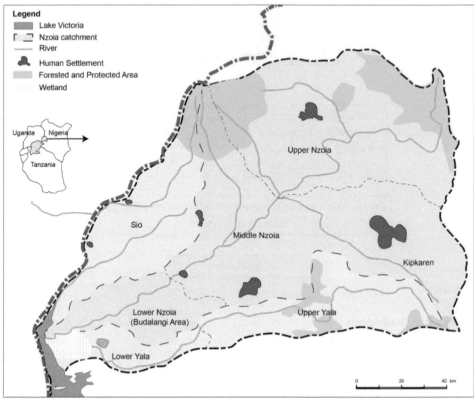

Figure 5.2. Nzoia River Basin

The Nzoia river basin is divided into three sub-catchments: the Lower Nzoia, characterised as flat and swampy; the Middle Nzoia and the Upper Nzoia, characterised with hills and steep slopes. The major tributaries of the Nzoia River are: Koitogos (Sabwani), Moiben, Little Nzoia, Ewaso Rongai, Kibisi, Kipkaren and Kuywa. The climate is tropical-humid and the area experiences four distinct seasons. Nzoia catchment has two rainy periods per year, one from March to May, with long rains and a second one from October to December, with short rains associated with ITCZ (the Inter Tropical Convergence Zone). The mean annual rainfall varies from a minimum of 1076 mm in the lowland to a maximum of 2235mm in the highlands. Average annual volume of precipitation of the catchment is about $1740 \times 10^6 m^3$. The average temperature of the area varies from 16ºC in the upper catchment (highlands) to 28º C in the lower catchment (lower semi-arid areas).

The dominant land use in the river basin is agriculture and the main agriculture production of the area are corn, sorghum, millet, bananas, groundnuts, beans, potatoes, and cassava and cash crops are coffee, sugar cane, tea, wheat, rice, sunflower and horticultural crops (Githui et al, 2008). The river basin plays a large role in economic development at local and also at national level. Major problems and challenges in the basin are soil erosion and sedimentation, deforestation, flooding, and wetland degradation. The area located at the most downstream end of the catchment is, as previously mentioned the Budalangi area, which is the focus of the present study. Floods are frequent in the Budalangi area (WMO/MWRMD/APFM, 2004) and their impact is felt through loss of life, damage to property and agricultural/crop destruction.

This case study is data scarce area. The lower the accuracy in the data, the lesser the accuracy in the predictions, therefore in data scarce areas this can result in bad or poor vulnerability predictions. Consequently, the results of the two approaches chosen may prove which one is a more appropriate approach to be used by the decision makers in such cases.

5.4.2. Assessing the flood risk of Budalangi region using physically based modelling

There are many simulation models available for solving problems of unsteady or steady flow. In this present study, an unsteady flow analysis was carried out using the SOBEK 1D/2D tool, developed by Deltares. SOBEK 1D/2D couples one-dimensional (1D) hydraulic modelling of the river channel to a two-dimensional (2D) representation of the floodplains. The hydrodynamic 1D/2D simulation engine is based upon the optimum combination of a minimum connection search direct solver and the conjugate gradient method. It also uses a selector for the time step, which limits the computational time wherever this is feasible. Detailed numerical implementation of the solution of the Saint Venant flow equations in SOBEK 1D/2D is given in the technical user manual of Verwey, (2006).

Generally the damages by flooding are classified as damages which can be quantified as monetary losses (tangible) and the damages which cannot be evaluated quantitatively in economic terms (intangible). These damages may be direct or indirect depending upon the contact to the flooding.

Flood damage estimation methodologies are applied worldwide (Dutta et al., 2003). For example, in the United Kingdom a standard approach to flood damage assessment is used (developed in the mid 1970s). Since then continually refined, this approach is mandatory for local authorities and agencies wanting central government assistance with flood mitigation measures. In United States, U.S. Army Corps of Engineers (USACE) has developed its own guidelines for urban flood damage measurement, (USACE, 1988). The method is based on the US Water Resources Council's 1983 publication on 'Principles and Guidelines for Water and Related Land Resources Implementation Studies'. The approach adopted in the method is very comprehensive for estimation of damage to urban buildings and to agriculture. In Australia, authorities considered that is no standard approach and it is a little attempt to achieve standard approach. Flood damage estimation methodologies are applied as well in many countries in Europe (Forster et al., 2008). These approaches are useful in conducting cost-benefit analyses of the economic feasibility of flood control measures.

This paper uses the Forster et al., 2008, approach where the expected damage (ED) on agriculture was calculated using the following equation, which is modified from Forster et al., (2008).

$ED = MV * Y * A * DI$, where ED – estimation damage; MV – market value; Y – yield per unit area; A- area of cultivation; DI – damage impact factor.

The number of houses in the inundated area was calculated using the information on population density and average number of family member per household.

$NH = \frac{IA*PD}{FM}$; where NH – number of houses in inundated area; IA – inundated area; PD – population density; FM – average number of family per household.

In order to estimate the flood damage, the estimation of some flood parameters are needed: flow velocity, depth and duration at any given point, proper classification of damage categories considering nature of damage, establishment of relationship between flood parameters and damage for different damage categories.

Flood Inundation Modelling

In order to build the 1D/2D hydrodynamic model of the Budalangi River, in SOBEK, available topographical information from the Shuttle Radar Topography Mission (SRTM) at a resolution of 90m by 90m and sparse cross-section data were used. Hydrograph variations at the upstream boundaries of the model were provided by a calibrated hydrological SWAT model of the Nzoia catchment. Recorded water levels for Lake Victoria were used as downstream boundary conditions. The SWAT model used to provide the upstream boundary condition was the one originally built and described by Githui et al. (2008) and recalibrated by van Hoey (2008). The 1:200 years design flood determined by SWAT was routed downstream by the hydrodynamic SOBEK model and inundation extents were drawn. A 1 in 200 year return flood was recorded on Nzoia River on November 2008, and therefore the inundation extent produced by the model was compared with available aerial information captured by to the Advanced Land Imager (ALI) on NASA's earth observing-1 satellite on the 13th November 2008.
The results of the model, at the moment of the largest flood extent, for the 1:200 return flood period are represented in Figure 5.3.

Flood Damage Evaluation
Many flood damage assessment methods have been developed since 1945 (White, 1945).

However, quantifying the expected flood damage is very difficult because the impact of a flood is a function of many physical and behavioural factors. For the purposes of this paper, flood damage was assumed to be related only to the flood depth.

The Budalangi region is a poorly developed rural area whose main industry is agriculture. Consequently the main expected damages were anticipated to be on the agricultural sector and were calculated based on a formula developed by Forster et al., (2008). The main cash crops in the area are known to be sugarcane, maize and rice. These crops were used, with readily available yield and expected local market values, to calculate the potential losses due to floods as a result of the 200 year return period event. In addition, loss of property and the affected population were included in the damage estimation, however it is recognised that in excluding the calculation of damage in relation to velocity this estimation is significantly simplified.

5.4.3. Assessing flood vulnerability of Budalangi using a parametric method

As mentioned above the parametric method used in this chapter is the one developed in Chapter 3, which consists in determining a flood vulnerability index (FVI), based on four components of flood vulnerability: social, economic, environmental and physical and their interactions, which can affect the possible short term and long term damages.

The four components of the flood vulnerability have been linked with the factors of vulnerability: exposure, susceptibility and resilience (Bosher et al., 2007, Penning-Rowsell and Chatterton, 1977), See Chapter 2 and Chapter 3, Equation 3.1.

The indicators belonging to exposure and susceptibility are increasing the flood vulnerability index therefore they are placed at nominator; however the indicators belonging to resilience are decreasing the FVI, this is why they are placed at denominator.

The application of this formula for each component leads to four distinct FVI indices; FVI_{Social}, $FVI_{Economic}$, $FVI_{Environmental}$ and $FVI_{physical}$, which aggregates into:

$$\text{Total FVI} = \frac{\left(\dfrac{E*S}{R}\right)_{Social} + \left(\dfrac{E*S}{R}\right)_{Economic} + \left(\dfrac{E*S}{R}\right)_{Environmental} + \left(\dfrac{E*S}{R}\right)_{Physical}}{4} \qquad \text{Eq. 5.2}$$

The exposure can be understood as the intangible and material goods that are present at the location where floods can occur, such as: loss of photographs and negatives, loss of life, delays in formal education (Penning-Rowsell et al., 2005). The susceptibility relates to system characteristics, including the social context of flood damage formation (Begum et al., 2007) and can be i.e. poverty, people with special needs, education, and level of trust. Susceptibility is defined as the extent to which elements at risk (Messner & Meyer, 2006) within the system are exposed, which influences the chance of being harmed at times of hazardous floods. Resilience to flood damages can be considered only in places with past events, since the main focus is on the experiences encountered during and after floods (Cutter, 1996, Cutter et al., 2003, Pelling, 2003, Walker et al., 2004, Turner II et al., 2003). Resilience describes the ability of a system to preserve its basic roles and structures in a time of distress and disturbances. Indicators showing resilience are flood insurances, amount of investments, dikes and levees, storage capacities, etc.

There are in total 29 indicators identified to contribute to eq (5.2), each ones with their own unit of measure. Not all indicators need to be always used while evaluating the FVI of a region. They always are evaluated and the most representative are used for FVI. A comprehensive description of such indicators in case of floods in the Mekong delta can be found in Quang et al (2012).

After identifying the indicators, in order to use them in eq (5.2) they need to be normalised using a predefined minimum and maximum. In general the classical proportional normalization is used, which keeps the relative ratios in the normalized values of the indicators as they were before normalization. The indicators become dimensionless, but still keep their proportion.

The FVI of each of the social, economic, environmental and physical component is computed using Eq. 5.1. The results of each FVI component (social, economic, environmental and physical) are summed up in Eq. 5.2.

The main issue while computing the FVI is actually to determine these indicators. There are different sources for determining the values of the indicators, and these are in general statistical data stored by environmental agencies, water boards, UN overviews and annual data from city halls.

5.5. Results obtained when applying the two approaches

5.5.1. Physically based modelling approach

The SOBEK simulation of the 1:200 year event results were water depths and inundation extents, as can be seen in Figure 5.2. The model is able to produce velocities of flow during an inundation event as well; however these velocities were not considered in the estimation of the damages and therefore not reported herein.

The maximum inundation extent was checked with an available satellite image on 13 November 2008. The obtained maximum inundation extent from the model was of 12.61km^2, which represents 97% of the inundation extent of the satellite image. Due to lack of data in the area, it is considered that this is good for the calibration of the model.

In order to determine the impact of flood and to evaluate the damages water depths obtained from the model were analysed. The obtained water depths were overall less than 2m (95% of the inundated area), and only 5% bigger than 2m in the upstream of the river. The main water depth is less than 0.5 m for 30% of the inundated area; 0.5m for 20% of the inundated area, between 1m and 1.5m for 35% of the inundated area; and 1.5 -2m for 10% of the inundated area.

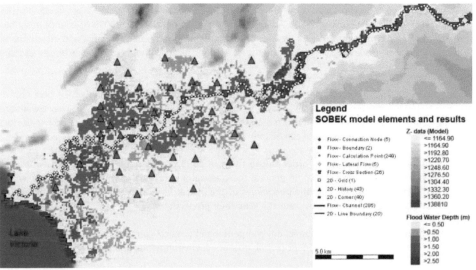

Figure 5.3. Lower Nzoia Flood Inundation Extent 1:200 year prediction

Based on the results from the hydrodynamic model, damage in the Budalangi area was computed using Forster et al, (2008) method and damage functions (Duggal & Soni, 2005).

In the Budalangi area the expected potential damages of 1.54M Euros (+/-80000 Euros was calculated for the event of 1:200 year return.

5.5.2. Parametric approach

The FVI methodology was applied to Budalangi Settlement, the results can be seen in Table 2. Budalangi vulnerability in the social and economic components is higher than the environmental and physical component, (1.00 means the highest vulnerability, see Table 1 for flood vulnerability index designations).

The data for the Budalangi area consulted to gather the indicators are: UNDP: United Nations Development Programme (HDI, child mortality, inequality); INTUTE: a web-site which provides social data for education and science research, (population density, unemployment, disabled people); the World Fact-Book, a database developed by the CIA with basic information on all the countries in the world (communication penetration rate, past experience); UNEDRA: University Network for Disaster Risk Reduction in Africa; Nzoia River Basin Management Initiative a public private partnership between Water Management Resource Authority and Mumia Sugar, Pan Paper and Nzoia Sugar Company (land use, flood insurance, shelters, closeness to river); DEFRA - Department for Environment, Food and Rural Affairs economic and statistical database at no cost charge (urban growth, population growth, amount of investment, dikes-levees, storage capacity); WKCDD & FMP, Western Kenya Community Driven Development & Flood Mitigation Project (river discharge, rainfall, evaporation); Western Water Board, Kenya (drainage, topography, industries, evacuation roads).

Table 5.1. Flood Vulnerability Designations

Designation	Index Value	Description
Very small vulnerability to floods	<0.01	Very small Vulnerability to floods, the area recover fast, flood insurances exist, Amount of investment in the area is high
Small vulnerability to floods	0.01 to 0.25	Social, economic, environmental and physical the area can once in a while experience floods, the area is vulnerable to floods and the recovery process is fast due to the high resilience measures, high budget, on the other hand if the area is less developed economic, even if a flood occurs the damages are not high, so small vulnerability to floods
Vulnerable to floods	0.25 to 0.50	Social, economic, environmental and physical the area is vulnerable to floods, the area can recover in months average resilience process, amount of investments is enough
High Vulnerability to floods	0.50 to 0.75	Social, economic, environmental and physical the area is vulnerable to floods, recovery process is very slow, low resilience, no institutional organizations
Very high vulnerability to floods	0.75 to 1	Social, economic, environmental and physical the area is very vulnerable to floods, the recovery process very slow. The area would recover in years. Budget is scarce.

Table 5.2. Budalangi FVI results

Budalangi Flood Vulnerability Index		
FVI Components	**FVI Values**	**FVI designation**
FVI Social	0.768	Very high vulnerability to floods
FVI Economic	0.521	High vulnerability to floods
FVI Environmental	0.314	Vulnerable to floods
FVI Physical	0.341	Vulnerable to floods
FVI Total	0.490	Vulnerable to High vulnerability to floods

Socially, the Budalangi area has very high vulnerability to floods, since has high population density, high child mortality rate, and a large affected population due to floods. The study also shows that the region has few shelters (0.6/km^2), no warning systems, no evacuation roads (no asphalted road), and only limited emergency services.

Economically the region is high vulnerable to floods since the area has low exposure to floods as the main economic activity is agriculture. The Human Development Index is low, and the area is not covered by flood insurance. Budalangi has few industries, the investment levels and a recovery process take long to recover after a flood event.

Environmentally, the Budalangi settlement is vulnerable to floods. The environmental component includes indicators which refer to damage to the environment caused by flood events or manmade interferences which could increase the vulnerability of certain areas. But activities like industrialisation, agriculture, urbanisation, deforestation, can increase the flood vulnerability, which may also create even more environmental damages.

When examining the physical vulnerability, the Budalangi area has very low slope and the settlement area is in contact with the river all along the length of the river so the exposure of Budalangi is high and has low resilience with little or no installed storage capacity.

Overall, the area following the designations of FVI is high vulnerable to floods, the recovery process is slow; the area has low resilience and no institutional organizations.

5.6. Discussion (Comparison – analysis and discussion of the approaches)

5.6.1 The physically based modelling approach

Physically based models have the advantage that they calculate the solution of a complicated and coupled set of equations that describe the phenomena of river flow and flooding. These models are dependent on physical knowledge that they incorporate into the equations and associated parameters. A key element for a good physically based model is the minimum of historical data that they need to determine the values for the parameters included in the physically based equations. Often, historical data is not available, in particular in areas of weak infrastructure, and this would make physically based models unusable in certain areas.

The advantage in using physically based models is their high capability for prognosis and forecasting, and their disadvantage is the high input data demand. In the past computational demand was a big disadvantage, but nowadays with the development of cloud and cluster computing capabilities over the internet, this disadvantage is reduced. However this is only true in case of larger, better-funded organisations that have good computer power to create cluster of computers, and not yet true for small consultancy companies or water boards who cannot dedicate cluster of computers for a specific modelling task. Due to the high computation resources demands, in case of 2D and 3D models, the calibration of physically or semi-physically based models can still be a tremendous effort.

In the present study the data on flooding was scarce, however the 2D physically based model was able to predict well the extent of flood, which shows that even in an ungauged catchment if the model is properly build, confidence in the construction of such a model does not require calibration (Cunge et al, 1980) and the results are good for design. A model based on the physics of the phenomena can be used to produce synthetic data to be used with a simple forecasting model.

One of the important tasks of the decision makers in flood situations is not only to take management decision but also to properly disseminate knowledge to involved stakeholders, including the general public. The objectives of knowledge dissemination is to offer simple and clear information, which can prepare the public for the future and also can actively involve the stakeholders in flood management planning. The information should be delivered in relevant spatial and temporal scales. A physically based model has the advantage that can offer all types of information on a very fine spatial resolution, at a level of a street, or a house, in a familiar and easily recognisable user interface. It is very important that the decision makers use thoroughly verified results, rather than results characterised by uncertainties, because the stakeholders and the public are taking often quick evacuation measures based on such information.

5.6.2 The parametric approach

The FVI approach regarding the *information on the mechanism and cause of flooding* has some limitations, what is given from this approach are the indicators values for river discharge, topography, closeness to the river, the amount of rainfall, dikes and levees. Considering these indicators the FVI approach can only evaluate the flood vulnerability, cannot tell the extent of flooding or the expected inundation area through the physical and environmental component. The application of this approach takes less preparation time than physically-based model construction, calibration and simulation.

The FVI approach regarding the *information on the health and safety implications to the affected population* is well designed; the approach shows through the social vulnerability indicators the exact population exposed to floods, the ones which are susceptible (youngest and eldest), if these people are aware and prepare, if they have and know how to interpret a warning system, which of the roads can act as an evacuation road. The social flood vulnerability index expresses whether the population of that specific area has experienced floods, the number of people working in the emergency service and the number and locations of shelters in the area. The social FVI provides a greater understanding of how people might be affected, which can feed into emergency services and evacuation strategy development.

The FVI approach regarding the *information on the economic damages and losses to the affected areas* gives basic damage estimation. The economic component is related to income or issues which are inherent to economics that are predisposed to be affected (Gallopin 2006).

Many economic activities can be affected by flooding events, among them are agriculture, fisheries, navigation, power production, industries, etc. The loss of these activities can influence the economic prosperity of a community, region or a country. The FVI can assess the economic vulnerability using a single number, though this number cannot evaluate the exact damage and losses but instead the index shows the number of industries in the area and their closeness to the river and also the amount of investment in counter measures and the number of flood insurances in that specific area.

How easily the information of the FVI approach is communicated?

From experts undertaking the study to the decision makers it can be said that the use of the FVI approach improves the decision-making process by identifying the vulnerability of flood prone areas. The FVI approach will direct decision-makers to a simplified usage and simpler understanding of the vulnerability; the FVI approach can be seen as a tool for decision making to direct investments to the most appropriate sectors and also to help in the decision-making process relating to flood defence, policies, measures and activities. The FVI approach allows, irrespective of uncertainties, relative comparisons to be made between case studies. While a level of uncertainty is inherent in FVI, the use of it in operational flood management is highly relevant for policy and decision makers in terms of starting adaptation plans. It may also be considered as a means to steer flood management policy in a more sustainable direction. However, as individual information is lost in the aggregation process, it needs to be retrieved by a more in-depth analysis of each process in order to design policies and their implementation.

From decision maker to the public:

Hence it is useful to have an *easy-to-apply* and *communicating* instrument that can help give an overview of the main points by having one single and comparable number, the FVI. The FVI is necessary, but not sufficient, for decision making and therefore should be used in combination with other decision-making tools. This should specifically include participatory methods with the population of areas identified as vulnerable and should also include a team of multidisciplinary thematic specialists and knowledgeable societal representatives and those with expert judgments.

5.7. Conclusions

The two approaches, modelling and index-based, have been applied to a data-scarce area - the Budalangi settlement. Examining the **approaches** from this study leads to the following conclusions:
1. It is clear that the FVI is not assessing directly flood risk, but has a contribution in assessing the risk; flood risk relates to "human health, the environment, cultural heritage and economic activity" (Scottish Government, 2009) since vulnerability takes a step further and covers some other aspects, such as:

- social (relates to two factors: on the one hand the presence of human beings which encompasses issues related to, for example, deficiencies in mobility of human beings associated with gender, age, or disabilities; on the other hand floods can destroy houses, disrupt communication networks, or even kill people. Included in this component are the administrative arrangements of the society, consisting of institutions, organizations and authorities at their respective level),
- environmental (deforestation, urbanization and industrialization have enhanced environmental degradation) and
- physical (relates to the predisposition of infrastructure to be damaged by a flooding event).

2. The parametric approach, here the FVI, through indicators is the only one which evaluates vulnerability to floods; the deterministic approach has a better science base, but limited evaluation of vulnerability;
3. FVI gives a wider evaluation, but is less rigorous. Therefore FVI is useful in a larger-scale vulnerability assessment, but a deterministic approach is better for more focused studies. In fact FVI could be used to decide where a deterministic model is necessary.

The Flood Vulnerability Index as analysed in the research provides a quick, reliable evaluation of flood vulnerability and in fact is the only method for assessing the vulnerability to flooding of a particular geographical area. The fact that indicators are calculated and used, allows for comparison of flood vulnerability in different areas as well as the identification of which indicators can determine the relative level of flood vulnerability. FVI can measure trends in the changing natural and human environments, helping identify and monitor priorities for action. These features, alongside the ability to identify the root causes of increased vulnerability, provide key information at a strategic level for flood risk planning and management. However the results would provide neither sufficient information nor the required level of detail for input into engineering designs or project level decisions.
The complex developments and dynamics in systems are not easy to include in the models.

FVI can provide an insight into the most vulnerable locations. It can analyse the complex interrelation among a number of varied indicators and their combined effect in reducing or increasing flood vulnerability in a specified location. It is very useful when there is a large level of uncertainty and decision makers are faced with a wide array of possible actions that could be taken in different scenarios, in this case the FVI can present easy to understand and to communicate results that would assist decision-makers in identifying the most corrective/effective measures to be taken. In this way proposed measures can be prioritised for areas that are at greatest risk. On the other hand this complexity is a negative point as wells, since it takes a long time and good knowledge of the area and the system behind the FVI to be able to implement it.

However, as all with models, this FVI model is a simplification of reality and its application should be compensated with thorough knowledge and expert-based analysis. The difficulties that the quantification of social indicators, as well as availability of other indicators poses to the calculation may constitute a considerable weakness of the model. The FVI is a useful tool to identify the most vulnerable elements of the water resource system and safety chain components (Pro-action, prevention, preparation, response and recovery).

FVI is a planning tool for risk assessment - the FVI represents a probability, as such it is inherently a tool to address uncertainty. At the same time it can give a false impression of certainty and it can be questioned whether a small set of indices for any river system for example, really contains any valuable information whatsoever.

Obviously such a parametric model is limited by the accuracy and availability of good datasets. A number of the indicators are very hard to quantify especially when it comes to the social indicators. On the other hand, such a model can give a simplified way of characterising what in reality is a very complex system, but seems to be the only one which assesses vulnerability. Such results will help to give an indication of whether a system is resilient, susceptible or exposed to flooding risks and help identify which measures would reap the best return on investment under a changing climate and population and development expansion. The important point is that such a model is used as one tool among others within the whole process of deciding on a roadmap for flood assessment.

CHAPTER 6

A Flood Vulnerability Index for Coastal Cities and its Use in assessing Climate Change Impacts

This chapter has been published as:

Balica, S.F., Wright, N.G., van der Meulen, F., 2012, A flood vulnerability index for coastal cities and its use in assessing climate change impacts, Natural Hazards, Springer Publisher, accepted on 17th May 2012

Dinh, Q., Balica, S.F., Popescu, I., Jonoski, A., (2012), Climate change impact on flood hazard, vulnerability and risk of the Long Xuyen Quadrangle in the Mekong Delta, International Journal of River Basin Management, 10:1, 103-120

6.1. Introduction

This chapter focuses on developing a Coastal City Flood Vulnerability Index (CCFVI) based on the same approach as the FVI methodology but for coastal flooding. It is well known that coastal cities are the most densely populated areas. The coastal cities and their inhabitants are exposed to high tidal waves, high storm surges, sea level rise, coastal erosion, etc. Therefore flood vulnerability assessments to coastal cities are needed. The chapter discusses in Section 6.2 the coastal system and coastal flood vulnerability, as well as the system approach related to coastal urban areas. Section 6.3 comprises the development of CCFVI, while in Section 6.4 comprises its application to nine cities around the world, each with different kinds of exposure. With the aid of this index, it is demonstrated which cities are most vulnerable to coastal flooding with regard to the system's components (see Section 6.2) that is, hydro-geological, socio-economic and politico-administrative. The index gives a number from 0 to 1, indicating comparatively low or high coastal flood vulnerability which shows which cities are most in need of further, more detailed investigation for decision makers.

Once its use to compare the vulnerability of a range of cities under current conditions has been demonstrated, it is used to study in Section 6.5 the impact of climate change on the vulnerability of these cities over a longer timescale. Section 6.6 is a discussion section on how to manage coastal cities through the CCFVI tool (www.unesco-ihe-fvi.org). The chapter ends with Section 6.7 where it is concluded that the use of CCFVI and climate change scenarios offer the opportunity to get a broad overview on components affected and on possible adaptation options that could be applied, directing resources at more in-depth investigation of the most promising adaptation strategies.

6.2. Background

Worldwide there is a need to enhance our understanding of vulnerability and to also develop methodologies and tools to assess vulnerability. One of the most important goals of assessing coastal flood vulnerability, in particular, is to create a readily understandable link between the theoretical concepts of flood vulnerability and the day-to-day decision-making process and to encapsulate this link in an easily accessible tool.

Coasts are highly dynamic and geo-morphologically complex systems, which respond in various ways to extreme weather events. Coastal floods are regarded as amongst the most dangerous and harmful of natural disasters (Douben, 2006). It is well known that the urban areas adjacent to the shorelines are associated with large and growing concentrations of human population, settlements and socio-economic activities. Considering the fact that 21 percent of the world's population lives within coastal zones (Gommes et al., 1997, Brooks et al., 2006), the potential impacts of sea level rise are significant for the wider coastal ecosystem (Kumar, 2006). Hoozemans 1993 carried out a global vulnerability assessment and estimated that, under current conditions, an average of 46 million people per year experience storm-surge flooding. Baarse (1995) suggests that some 189 million people presently live below the one-in-a-hundred-year storm-surge level. There is therefore a need for a readily calculated and easily understood method to calculate flood vulnerability is such areas.

This work is built on earlier works on a flood vulnerability index in river basins (see Chapter 2) to establish a flood vulnerability index using a composite method. This index can then be used to identify the most vulnerable coastal cities, develop adaptation measures for them and assess the effects of future change scenarios.

This chapter focuses on developing a Coastal City Flood Vulnerability Index (CCFVI) based on exposure, susceptibility and resilience to coastal flooding. The chapter is focused on large cities in low-lying deltaic environments with soft sedimentary coasts (estuaries, lagoons, mangroves, dunes, beaches). These cities experience both the influence of river discharge and of the sea and they are, by consequence, very vulnerable to impacts of climate change. It is applied to nine cities around the world, each with different kinds of exposure. With the aid of this index, it is demonstrated which cities are most vulnerable to coastal flooding with regard to the system's components, that is, hydro-geological, socio-economic and politico-administrative. The index as the FVI gives a number from 0 to 1, indicating comparatively low or high coastal flood vulnerability which shows which cities are most in need of further, more detailed investigation for decision makers.

Once its use to compare the vulnerability of a range of cities under current conditions has been demonstrated, it is used to study the impact of climate change on the vulnerability of these cities over a longer timescale.

The results show that CCFVI provides a means of obtaining a broad overview of flood vulnerability and the effect of possible adaptation options. This, in turn, will allow for the direction of resources to more in-depth investigation of the most promising strategies.

6.3. Defining coastal system and coastal flood vulnerability

It is expected that, due to climate change, coastal communities around the world will be increasingly affected by floods. In fact, some are already considered vulnerable to ongoing climatic variability (IPCC, 2007; MIZRA, 2003). Climate change is expected to cause accelerated sea level rise with elevated tidal inundation, increased flood frequency, accelerated erosion, rising water tables, increased saltwater intrusion, increasing storm surges and increasing frequency of cyclones (Fenster & Dolan, 1996). Apart from this, population growth and increasing urbanisation cause marine and coastal degradation (UNEP, GEO-3, 2002).

6.3.1. What is a coastal system?

Coasts are dynamic systems, undergoing adjustments of form and process (termed morph dynamics) at different time and space scales in response to geo-morphological and oceanographically factors (Cowell et al., 2003a, b). Human activity exerts additional pressures that may dominate over natural processes. Coastal landforms, affected by short-term perturbations such as storms, generally return to their pre-disturbance morphology, implying a simple, morphodynamic equilibrium (Woodroffe, 2003, Crooks, 2004).

The natural variability of coasts can make it difficult to identify the impacts of climate change. For example, most beaches worldwide show evidence of recent erosion, but sea-level rise is not necessarily the primary driver. Erosion can result from other factors, such as altered wind patterns (Pirazzoli et al.,

2004; Regnauld et al., 2004), offshore bathymetric changes (Cooper and Navas, 2004), or reduced fluvial sediment input (Nicholls et al., 2007), or hard structures built near the coast.

Natural coastal systems

The IPCC, (2007) distinguishes between the following natural coastal systems: deltas, estuaries and lagoons, beaches, rocky shorelines and cliffed coasts, mangroves, sea grasses and coral reefs. In this chapter, we focus on large urban areas situated in deltas. A delta is an area where the river sediment is building out into the sea. Deltas are biologically rich and diverse systems with waterfowls, fish and vegetation and they support a large economic system based on tourism, agriculture, hunting, fishing, harbour and industry development (EC-JRC, 2005; Prakasa, 2005). Consequently, deltas are often densely populated (Ericson et al, 2006. IPCC, 2007). Many people in deltas are already subject to flooding from both storm surges and seasonal river floods, and therefore it is necessary to develop further methods to assess coastal cities flood vulnerability.

6.3.2. Coastal flood vulnerability

Large populations are found in coastal areas where the exposure to coastal floods is high (Small and Nicholls, 2003). The number of people affected is likely to increase, due to net coastward migration across the globe (Bijlsma et al., 1996). On the one hand, some of the exposed populations are protected from flooding by various structural and non-structural measures that are part of the resilience strategy. On the other hand, some of them have none, or only weak, flood defences and the exposed populations are more often subject to flooding with the consequent disruption, economic loss, and loss of life. Smith and Ward 1998 showed that rising sea levels will raise flood levels, it is also estimated that the number of people flooded in a typical year by storm surges would increase 6-times and 14-times given a 0.5 and 1.0m rise in global sea levels respectively (Nicholls, 2004).

Coastal vulnerability indices

Coastal vulnerability indices have been developed as a rapid and consistent method for characterising the relative vulnerability of different coastal areas. The simplest of these are assessments of the physical vulnerability of the area, while the more complex also examine aspects of economic and social vulnerability. A summary of various indices applied globally is given here.

In previous work on coastal vulnerability, approaches were derived from Gornitz, 1991, Gornitz and Kanciruk (1989), Thieler, (2000) with an index widely applied in the United States and in a modified form in Canada and parts of South Africa. It has also been viewed as important to incorporate social data on people at risk, the most detailed social vulnerability analysis being the synthesis by Boruff et al. (2005), Abuodha and Woodroffe, (2007), Gornitz, (1994)). The social vulnerability index (SoVI) uses "initially 42 socio-economic variables, reduced to 11 statistically independent factors" (i.e. age, race, ethnicity, education, family structure, social dependence, occupation) (Cutter et al, 2003). This being applied on a coastal county, basis in a principal component analysis (PCA) to produce the overall coastal social vulnerability score (CSoVI). The coastal social vulnerability score (CSoVI) is a combination of variables for North America and Australia applied specifically to the beaches. The variables are: dune height, barrier type, beach type, relative sea-level change, shoreline erosion and accretion, mean tidal range and mean wave height. In 2010, McLaughlin and Cooper developed a multi-scale coastal vulnerability index to investigate the implications of spatial scale in depicting coastal hazard risk, coastal vulnerabilities for national, local authority and site level. The authors in this index referred to coastal erosion vulnerability, either than coastal flood vulnerability. This can be seen in the variables which were used: a coastal characteristics sub-index concerned with the resilience and susceptibility of the coast to erosion, a coastal forcing sub-index to characterize the forcing variables contributing to wave-induced erosion and a socio-economic sub-index to assess the infrastructure potentially at risk. The socio-economic sub-index (McLaughlin and Cooper, 2010) comprises 6 variables: population, roads (vital lines of communication and transport), cultural heritage, railways, land use and conservation status). Sharples, (2006) uses identification and mapping of coastal substrates and landforms (i.e. geomorphic types) in order to assess coastal vulnerability which has greater or lesser sensitivity to potential coastal impacts of climate change and sea level rise, such as accelerated erosion and shoreline recession, increased slumping or rock fall hazards, changing dune mobility and other hazards.

6.3.3. Components of coastal flood vulnerability – the systems approach

The systems approach aims to identify the interactions of different actors or components within certain defined boundaries. Van Beek (2006) identifies three interdependent subsystems in the coastal vulnerability system:

- The **natural river subsystem** (NS), in which the physical, chemical and biological processes take place;
- The **socio-economic subsystem** (SES), which includes the societal (human) activities related to the use of the natural river system; Socio-economic systems are made up of rules and institutions that mediate human use of resources as well as systems of knowledge and ethics that interpret natural systems from a human perspective (Berkes and Folke, 1998, Adger, 2006).
- The **administrative and institutional subsystem**, that includes administration, legislation and regulation, where the decision, planning and management processes take place.

Each of the three subsystems is characterised by its own elements and it is surrounded within its own environment. In this chapter, the vulnerability system is the coastal city. It can be seen as a set of interconnecting systems; the system is composed of interacting elements where different processes are carried out using various types of resources. In this context, one must define the system through its components and interactions. It should be shown how each element of the system, as well as the individual interactions, are vulnerable. For a better understanding of this chapter, the delineation between terminologies is presented.

Hazards have many origins, but in this chapter, we normally view them as caused by the interaction between society and natural systems (e.g. Precipitations, Floods, and Cyclones). Unexpected hazards become visible rapidly such as flooding or hurricanes, and last for a small period ranging from hours to weeks. Continuous hazards are very slow events that are barely perceptible by society such as sea level rise. This chapter considers both types of hazards. From the perspective of this research, vulnerability is embedded as a combination of the susceptibility of a given population, system, or place to harm from exposure to the hazard and directly affects the ability to prepare for, respond to, and recover from hazards and disasters. Resilience speaks to the capacity of the population, system, or place to buffer or adapt to changing hazard exposures.

The natural coastal system is delimited by climate and (hydro-geo) physical conditions (catchment and coast), the socio-economic system is formed by the demographic, social and economic conditions of the surrounding economies and the administrative and institutional system is formed and bounded by the constitutional, legal and political system. Coastal floods distress three components of the coastal vulnerability system, each of them belong to one of the subsystems described here, and their interactions affect the possible short-term and long-term damages. The components can be assessed by different indicators to understand the vulnerability of the system to coastal floods. The system components are: hydro-geological, socio-economical and politico-administrative.

6.3.3.1. Hydro-geological Component

The hydro-geological component is a part of the natural coastal system, being hazard dependent. It comprises hydro-geo-morphological (i.e. sea level rise, river discharge, soil subsidence) and climatic (i.e. number of cyclones, storm surge) characteristics of the system. This component uses only the exposure indicators, see eq. 3, because the system, here the city, is characterised from hydro-geological point of view (exposure), the physical part, as infrastructure (resilience) are taken into consideration under the socio-economic and politico-administrative component. The hydro-geological component continues to relate to the interrelation between the environment and the vulnerability associated with this interaction (Villagran, 2006). Developments such as land subsidence, storm surge, and high river discharge have enhanced environmental degradation, aggravating effects of climate change and associated sea level rise, increasing the potential occurrence of floods. In this case, the hydro-geological component does not consider urbanization and industrialization as hazard.

6.3.3.2. Socio-economic Component

The socio-economic component is part of the socio-economic system; the flooding affects the day-to-day lives of the population that belongs to the system. The socio component relates to the presence of human beings and encompasses issues related to it e.g. deficiencies in mobility of human beings associated with gender, age, or disabilities. Coastal floods can cause destruction of houses, disruption in communications, in the agricultural process, or even fatalities. The economic component is related to income or issues which are inherent to economics that are predisposed to being affected. There are many economic activities, which can be negatively affected by coastal flooding. Among them are tourism, fisheries, navigation, industries, agriculture, and availability of potable water etc. This influences the economic prosperity of a community, region, urban area or a country. McLaughlin and Cooper, 2010, agreed that the choice of socio-economic variables adds an inherent cultural bias to an index.

6.3.3.3. Politico-Administrative Component

The politico-administrative component encompasses the administrative and institutional system. To characterise the administrative and institutional system, the relevant institutions at the national, regional and local level have to be identified. The approach assumes that one or more institutions have the ability and authority to develop and implement plans that will oversee and manage the coordinated development and operation of the actions of the local authorities that affect the coasts. This component embraces the exposure, susceptibility and resilience indicators see eq. 6.6

The components can be assessed by using different indicators. These components have been linked with the three factors of vulnerability. The index aims to describe flood damage at coastal city level. Consequently, through this identification decision makers can make informed choices on how to best allocate resources to ameliorate flood damage in the future.

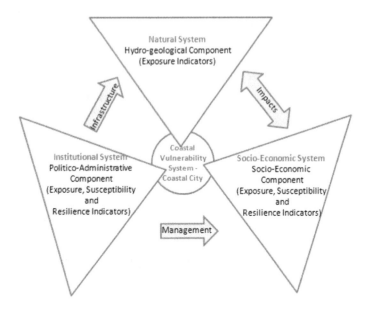

Figure 6.1 Coastal vulnerability system: its sub-systems and interactions (modified from van Beek, 2006)

The relationship between the components affected by coastal floods and the subsystems of the coastal vulnerability system are shown in the Figure 6.1

Understanding the natural processes as well as the economic and social services or functions that coasts fulfil is critical to the successful and sustainable management of these systems.

Considering the coastal system, the NS gives/receives impacts to/from the SES by i.e. the impact of sea level rise is obvious: more severe storms; while storm surge will decrease a little because of the larger water depths, it will also increase because of the more severe storm activity; tidal prisms will increase, etc. As results, salt intrusion will increase, structures will be more stressed, wetlands will be inundated and disappear (Covich, 1993). All of these will affect the population along the coasts, which is permanently increasing (Hutchings and Collett, 1977). Damage will also occur to agricultural areas because of the additional saltwater intrusion. The cultural heritage will be susceptible to flood (for example: Venice and the Netherlands).

The coast is affected by human activities such as bank protection, shipping, and construction and operation of hydraulic infrastructures. The coasts provide tangible and direct economic benefits. Tourism, fisheries prosper on the wealth of natural resources coasts supply. The protected coastal waters also support important public infrastructure, serving as harbours and ports vital for shipping, transportation, and industry. To maintain and enhance these and other services and benefits derived from coasts, they must be managed.

6.4. Development of Coastal City Flood Vulnerability index methodology

6.4.1. Flood Vulnerability Factors

Societies are vulnerable to floods due to three main factors; exposure, susceptibility and resilience. The vulnerability of any system (at any scale) is reflective of (or a function of) the exposure and susceptibility of that system to hazardous conditions and the resilience of the system to adapt and/or recover from the effects of those conditions (Smit & Wandel, 2006), See Figure 6.2 for the coastal city system. See the factors' definitions on Chapter 2, Section 2.3.1

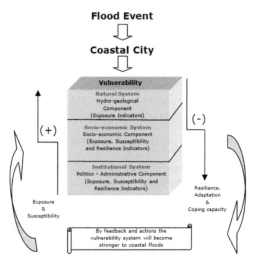

Figure 6.2 Vulnerability System

Based on Le Chatelier's principle (Hatta, Tatsuo, 1987) "Any change in status quo prompts an opposing reaction in the responding system". All systems are in hazard, but their vulnerability reflects the possible damage which can be expected in the case of an event. All the components of the system can be affected by floods. After each flooding event, the social system usually becomes stronger and gives feedback to reduce the vulnerability to future floods. A system at risk is more vulnerable when it is more exposed to a hazard and the more it is susceptible to its forces and impacts. However, the system will be less vulnerable when it is more resilient and less exposed.

6.4.2. Coastal vulnerability indicators

Vulnerability in considered here based on the use of indicators (see indicator's definition on Chapter 2, Section 2.2.1). Therefore is very important to know the impacts on the people, cities, natural resources, via the use of these indicators.

McLaughlin and Cooper, 2010, affirm that it is tempting to use all available data in the creation of an index. Previously, Dal Cin and Simeoni, 1989, claimed that using more variables gives results that are more correct. This, however, is not necessarily true since variables/indicators can be highly correlated.

Since the development of the FVI involves the understanding of different relational situations and characteristics of a coastal system exposed to flood risks, a deductive approach to identify the best possible indicators has been used, based on the principles and the conceptual framework of vulnerability outlined in the previous chapters.

Understanding each concept and considering certain indicators may help to characterise the vulnerability of different systems, by which actions can be identified to decrease it. Every vulnerability factor (exposure, susceptibility and resilience) represents a set of constituent indicators, based on the characteristics of a coastal system, which can help to better understand the response of the coastal cities to floods.

Some of the indicators, in general, belong to two of the factors. Due to definition used in this study, the indicators are considered only for one of the factors. For example, Flood Protection (FP) indicator belongs to resilience factor, as being a positive measure; the method does not include this Environmental Component where the FP indicator can be a negative measure. The Institutional Organisations (IO) indicator is used to index the politico-administrative component, from the functional relationship with the vulnerability. In (Table 6.4) it can be seen that higher number of IO, lower the vulnerability. The IO can be included under the susceptibility factor as well, but because of the difficulty in quantifying corruption, this indicator is not considered; the choice is to use it only as resilience indicator.

Table 6.1 Relationship between components and factors

<table>
<tr><td colspan="9" align="center">Coastal Flood Vulnerability Indicators</td></tr>
<tr><td rowspan="2">Water Resources System</td><td rowspan="2">Coastal Flood Vulnerability Components</td><td colspan="7" align="center">Vulnerability Factors</td></tr>
<tr><td>Abb</td><td>Exposure</td><td>Abb</td><td>Susceptibility</td><td>Abb</td><td>Resilience</td></tr>
<tr><td rowspan="7">Natural</td><td rowspan="7">Hydro-geological</td><td>SLR</td><td>Sea level rise</td><td rowspan="7" align="center">The Hydro-Geological Component doesn't consider susceptibility indicators for this method</td><td></td><td rowspan="7" align="center">The Hydro-Geological Component doesn't consider resilience indicators for this method</td><td></td></tr>
<tr><td>SS</td><td>Storm surge</td><td></td><td></td></tr>
<tr><td># Cyc</td><td># of cyclones in the last 5 years</td><td></td><td></td></tr>
<tr><td>RD</td><td>River Discharge</td><td></td><td></td></tr>
<tr><td>SF</td><td>Foreshore Slope</td><td></td><td></td></tr>
<tr><td>Soil</td><td>Soil Subsidence</td><td></td><td></td></tr>
<tr><td>CL</td><td>Km of coastline</td><td></td><td></td></tr>
<tr><td rowspan="2">Social</td><td rowspan="2">Social</td><td>PCL</td><td>Population close to CL</td><td>%Disable</td><td>% of disable persons</td><td>S</td><td>Shelters</td></tr>
<tr><td>CH</td><td>Cultural Heritage</td><td></td><td>(< 14 and > 65 years)</td><td>A/P</td><td>Awareness & Preparedness</td></tr>
<tr><td rowspan="2">Economic</td><td rowspan="2">Economic</td><td rowspan="2">GCP</td><td rowspan="2">Growing coastal population</td><td rowspan="2" align="center">The Economic Component doesn't consider susceptibility indicators for this method</td><td></td><td>Drain</td><td>Km of Drainage</td></tr>
<tr><td></td><td>RT</td><td>Recovery Time</td></tr>
<tr><td rowspan="2">Institutional</td><td rowspan="2">Politico-Administrative</td><td>UP</td><td>Uncontrolled Planning Zones</td><td>FHM</td><td>Flood Hazard Maps</td><td>IO</td><td>Institutional Organizations</td></tr>
<tr><td></td><td></td><td></td><td></td><td>FP</td><td>Flood Protection</td></tr>
</table>

The relation of coastal vulnerability components, indicators and factors is illustrated in Table 6.1. The availability of data, the importance of certain indicators and the condition that all FVI's computed must be dimensionless for the purposes of comparison, led to the formulation of the equations for each vulnerability component.

6.4.3. General flood vulnerability index equation related to coastal cities

A general FVI equation (Eq. 6.2) for all scales is described in Chapter 3. The equation links the values of all indicators to flood vulnerability components and factors (exposure, susceptibility and resilience), without weighting, (Cendrero and Fisher, 1997, Peduzzi et al., 2001, Briguglio, 1992, 1993, 1995, 1997, 2003). This is done because of different number of rating judgments which "lie behind combined weights", or interpolating.

McLaughlin and Cooper, 2010, use Gornitz (1990) approach, a scale of 1–5 is chosen, with 5 contributing most strongly to vulnerability and 1 contributing least. The 1–5 scale that was used for every variable standardises the scoring system and enables variables measured in different units to be combined mathematically.

McLaughlin and Cooper, (2010) use the approach of Gornitz and White (1992), which is based on the fact that the "sum of the variables was less sensitive than one based on the products of the variables". This research used the approach of FVI (Balica et al., 2009) (Eq. 6.1); the approach is based on the fact that each system has its own vulnerability to floods, so a variable cannot be considered zero.

The procedure for calculating the CCFVI starts by converting each identified indicator into a normalised (on a scale from 0 to 1), dimensionless number using predefined minimum and maximum values from the spatial elements under consideration. Equation (1) is showing the expression used for normalisation.

$$NV_i = \frac{RVi}{Max_{i=1,n}(RVI_i)}$$

6.1

where NVi represents the normalised value of the indicator I, the RVi represents the real value of the indicator I, and $Max_{i=1,n}(RVI_i)$ represents the maximum value from a set of n computed real values of the indicator I (where n is the number of spatial elements under consideration). Normalised indicators are subsequently used for CCFVI calculations.

The CCFVI of each coastal component (hydro-geological, social, economic and politico-administrative) is computed based on the general flood vulnerability index (FVI) formula (Eq 3.1, Chapter 3).

The general formula for FVI is computed by categorising the indicators to the factors to which they belong (exposure (E), susceptibility (S) and resilience (R)) (Cendrero & Fischer, 1997). The indicators of exposure and susceptibility are multiplied and then divided by the resilience indicators, because indicators representing exposure and susceptibility increase the flood vulnerability and are therefore are placed in the nominator. The resilience indicators decrease flood vulnerability and are thus part of the denominator.

The indicators play a gradually more significant policy role; also, they represent only synoptic sides of a system. The first step in an indicator-based vulnerability assessment is to select indicators (Adriaanse, 1993; World Bank, 1994, 1997; CRED, 2008; Perry, 2006; Quarantelli, 2005; Sorenson & Sorensen, 2006; Briguglio, 2003; Peduzzi et al., 2001; for example, the detailed World Bank Africa Database 2005, consists of almost 1200 indicators (World Bank, 2005). The benchmark is to gather a list of proxies using the following criteria: suitability, definitions or the theoretical structure, availability of data.

As will be presented in Tables 6.2-6.4, a total number of 19 indicators is used in general to assess the vulnerability of coastal areas. These 19 indicators were selected from World Bank data set, 2001, Gorintz, 1990, McLaughlin & Cooper, 2010 and Cutter et al., 2003, after using multi-collinearity analysis among 30 coastal indicators for 9 case studies.

Coastal City Flood vulnerability index for Hydro-Geological Component

$$FVI\text{Hydro-Geological} = f\{SLR, SS, \#Cyc, FS, RD, Soil, CL\}$$

6.2

Table 6.2 Indicators information of the hydro-geological component

Indicators	Abb.	Factor	Unit	Definition	Functional relationship with vulnerability
Sea Level Rise	SLR	Exposure	mm/year	How much the level of the sea is increasing in 1 year	higher SLR, higher vulnerability
Storm Surge	SS	Exposure	cm	A storm surge is the rapid rise in the water level surface produced by onshore hurricane winds and falling barometric pressure.	bigger increase in WL, higher vulnerability
# of Cyclones	#Cyc	Exposure	#	Number of cyclones in the last 10 years	higher # of Cyclones, higher vulnerability
River Discharge	RD	Exposure	m3/s	maximum discharge in record of the last 10 years, m³/s	Higher RD, higher vulnerability
Foreshore Slope	FS	Exposure	%	Foreshore Slope and depth of the sea near the coast, can change a lot and often. Average slope of the foreshore beach	Lower slope, higher vulnerability
Soil subsidence	Soil	Exposure	m2	How much the the area is decreasing?	Higher areas, higher vulnerability
Coastline	CL	Exposure	km	Kilometers of coastline along the city	longer CL, higher vulnerability

Coastal City Flood vulnerability index for Social and Economic Component

$$FVISocial = f\frac{CH,PCL,\%Disable}{A/P,S}$$

6.3

$$FVIEconomic = f\frac{GCP}{RT,Drainage}$$

6.4

Table 6.3 Indicators information of the socio-economic component

Indicators	Abb.	Factor	Unit	Definition	Functional relationship with vulnerability
Cultural Heritage	CH	Exposure	#	number of historical buildings, museums, etc., in danger when coastal flood occurs,	high # of CH, higher the vulnerability
Population close to coastline *	PCL	Exposure	people	Number of people exposed to coastal hazard	The higher number of people, higher vulnerability
Growing coastal population *	GCP	Exposure	%	% of growth of population in urban areas in the last 10 years	fast GCP, higher vulnerability, hypothesis is made that fast population growth may create pressing on land subsidence
Shelters	S	Susceptibility	#	number of shelters per km² including hospitals	bigger # of S, lower vulnerability
% of disable persons (<14 and >65)	%Disab	Susceptibility	%	% of population with any kind of disabilities, also people less 12 and more than 65	higher %,higher vulnerability
Awareness & Preparedness *	A/P	Resilience		Are the coastal people aware and prepare for floods? Did they experience any floods in the last 10 years?(Scaled)	Higher # of past floods, more prepare/aware, lower vulnerability
Recovery Time *	RT	Resilience	days	Amount of time needed by the city to recover to a functional operation after coastal flood events (Scaled)	the higher amount of time, the higher vulnerability
Km of Drainage	Drain	Resilience	Km	Km of canalization in the city	higher km, low vulnerability

Explanation of indicators: the indicator "population close to the coastline" is defined as the number of people exposed to coastal hazards. For example," population living in the flood prone area along the coast", "growing coastal population" refers to the percentage growth in population of the urban area which signifies the economic wealth of the urban area. The indicator "awareness/preparedness" was scaled between 1-10, where 1 is given for the area where population has no concern with floods and 10 for the urban area where the population has experienced floods for a long time. This indicator is aware of the potential floods in the area, i.e. they have trust in institutions to mitigate the harm of floods, they have flood insurance, they understand the consequences and restrictions of their actions towards flood protection and they are prepared for emergency situations.

The Drainage (D), Recovery Time (RT) and Growing Coastal Population (GCP) indicators belongs to the economic component (Eq. 6.4 and 6.6'), and they are associated with the economy state of an area because:

Drainage indicator represents the length of canals in the area (in kilometers) and belongs to resilience (i.e. will be seen in the denominator in Eq. 6.4. The indicator has an indirect relation with the vulnerability, longer kilometers of drainage, lower the vulnerability. The drainage indicator reflects the economic state of the area, higher number of kilometers of drainage, richer the region.

The RT indicates the amount of time needed by the city to recover to a functional operation after coastal flood events. Longer the time, higher the vulnerability. This indicator belongs to the economic component because it reflects the wealth of an area. Richer states recover faster, due to the higher GDP/capita for example. The RT indicator was scaled between 1-10, where 1 means all economic activities are strongly damaged and they may not recover for many years and 10 means that the economic activities of the region are hardly affected by floods, either in the short, or in the long term.

The GCP refers to the percentage growth in population in the urban area which shows the economic wealth of the urban area, and therefore belongs to exposure factor (i.e. will be seen at the nominator in Eq 6.4.). The GCP has a direct relation with the vulnerability, higher the GCP, higher the vulnerability. The GCP is an indicator, which reflects the economic state of an area.

Coastal City Flood vulnerability index for Politico-Administrative Component

$$\text{FVIPolitico-Administrative} = f\frac{FHM, UP}{IO, FP}$$ 6.5

Table 6.4 Indicators for the politico-administrative component

Indicators	Abb.	Factor	Unit	Definition	Functional relationship with vulnerability
Flood Hazard Maps	FRP	Susceptibility	_	Flood Hazard Mapping is a vital component for appropriate land use planning in flood-prone areas.	existance of those measures, lower vulnerability
Institutional Organizations	IO	Resilience	#	Existance of IO	Higher #, lower vulnerability
Uncontrolled Planning Zone	UP	Exposure	%	% of the surrounding coastal area (10 km from the shoreline) is uncontrolled	Higher %, higher vulnerability
Flood Protection	FP	Resilience	_	The existance of structural measures that physically prevent floods from entering into the city (Storage capacity)	if YES, lower vulnerability

Total Coastal City Flood Vulnerability Index

Total FVI = Hydro-Geological + Social + Economic + Politico-Administrative   6.6

$$\text{TotalFVI} = \left\{ (SLR, SS, \#Cyc, FS, RD, Soil, CL) + \left(\frac{CH, PCL, \%Disable}{S, A/P} \right) + \left(\frac{GCP}{RT, Drainage} \right) + \left(\frac{FHM, UP}{IO, FP} \right) \right\}$$ 6.6'

The integrated Coastal City Flood Vulnerability index is a method to combine multiple aspects of a system into one number. On a global perspective the results will be presented in values between 0 and 1; 1 being the highest vulnerability found in the samples studied and 0 the lowest vulnerability.

6.5. Application of the CCFVI methodology in order to assess FVI of different deltaic areas, using FVI tool

The CCFVI methodology uses a total of 19 indicators. Nine case studies were selected based on city size and different physiographic setting. They are: Buenos Aires (Argentina), Calcutta (India), Casablanca (Morocco), Dhaka (Bangladesh), Manila (Philippinnes), Marseille (France), Osaka (Japan), Shanghai (China) and Rotterdam (the Netherlands). The nine cities selected are in both developed and developing countries.

6.5.1. Data sources

The following data sources were used to assess the values of the flood vulnerability indicators for each of the cities. The data collection was done via the internet and so used readily available sources. An accurate assessment of flood vulnerability is difficult, due to the lack of necessary data and because vulnerability is geographically and socially differentiated (Adger, 2006). Many varied data sources were consulted, as seen in the following itemise list. Multidisciplinary data was used in the assessment of coastal flood vulnerability index, the accuracy of data potentially can edged vulnerability asssessment. Using varied sources with different units, add to this data process a carefull conversion in order to display the same data.

It is reckoned therefore that the overall ranking of the FVI assessment may be crude, however precise data are not available to support a more refined ranking evaluation. The focus should be on the approach, not the data availability.

1. Casablanca "Academie de l'eau", the website of Environmental Ministry of France and "Gouvernement du Royaume du Morocco", the website of Government of Morocco, and the International Federation of Red Cross for floods in Morocco.
2. For Calcutta various sources were consulted, the Department of Natural Resources, Weather Underground India, World FactBook, (database developed by the CIA with basic information on all the countries in the world), World Bank, UNICEF India, ADB Asian Development Bank, Megaessays (source for high quality essays on a wide range of subjects), Debasti et al, (1995), GIS Development of India, Ministry Water Resources India.
3. For Dhaka: World Bank, HighBeam research, World FactBook were used.
4. For Manila the website of PAGASA, Phillipinnes Athmosferic, Geophysical and Astronomical Services Administration, the article of Zoleta-Nantes, (1999), ICHARM and Connor & Hiroki (2005) were used.
5. Buenos Aires. AIACC, Assessments of Impacts and Adaptations to Climate Change, Academie de l'Eau, Bnamericas, (the leading source of business information in Latin America) were the sources.
6. For Osaka city: Kawai, (2008), Japan River Organisation, Kadoya et al., (1993), MLIT website (Ministry of Land, Industry and Transport), ICHARM.
7. For the city of Marseille the Academie de l'eau website, Marseille Municipality and the Water Resources eAtlas (an electronic Atlas developed by IUCN, IWMI, Ramsar and WRI) website were used.
8. For Shanghai, the Yangtze River Committee, UNESCAP data sources were taken into consideration.
9. For Rotterdam: UNESCAP, Walstra, (2009), Bijker, (1996), EUROSION, (Holland Coast) were used.

The results of the CCFVI's for all components and the total CCFVI, are summarised below and easily computed using the FVI tool, unesco-ihe-fvi.org (Balica & Wright, 2009) for coastal cities.

6.5.2. The Hydro-geological Component

The values of the hydro-geological component indicators were used in Equation 6.2, described in the section above; the results of the hydrogeological component are shown in Figure 6.3. Seven indicators are used to determine the hydrogeological CCFVI values. The indicators all belong to the exposure factor.

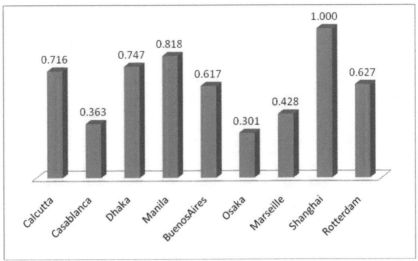

Figure 6.3 CCFVI Hydro-geological Component (exposure to coastal flood).

When examining the hydro-geological exposure, from the FVI it can be seen that Shanghai is the most exposed to coastal floods. This is mainly due to its high length of coastline and the high value of river discharge. Following in the FVI, the next vulnerable out of the assessed case studies is the city of Manila, largely due to its exposure to tropical cyclones and flooding. The recent tropical storm Ketsana (2009) illustrates this exposure of Manila and the surrounding area to environmental threats. With flood waters reaching nearly 7m a.s.l in some city areas (WWF, 2009) and hundreds of deaths during this one storm, Manila is highly vulnerable. Dhaka and Calcutta come next largely because of their storm surge, coastal line length and river discharge. Rotterdam city is situated on the fifth place, with a very high soil subsidence and a high Rhine river discharge. The city of Buenos Aires ranks the sixth, with a very high river discharge value but very low amount of kms of coastline, storm surge, number of cyclones. Marseille and Osaka are coming close in the rank as seven and eight. The least vulnerable from all the examples is Casablanca with small number of cyclones in the last 10 years, few kms along the coast and no river discharge However, this is not implying that the city is not vulnerable to coastal floods. All these cities have already been subjects of coastal floods with loss of life and significant damage costs.

6.5.3. The Social Component

The values of the social component indicators were used in Equation 6.4, described above. The results of the social component are shown in Figure 6.4. Five indicators, belonging to all factors of vulnerability, were used to determine the social CCFVI values. These were: population, cultural heritage, shelters, flood cultural behaviour (awareness and preparedness), % of disable population. Using these criteria, from the FVI point of view, Shanghai stands out as the most vulnerable to coastal floods, mainly due to its high number of people living in coastal flood prone areas, fewer shelters. The second most vulnerable city to coastal floods is Dhaka for similar reasons. Calcutta and Manila are third/fourth most vulnerable, while Osaka and Buenos Aires come next. The City of Rotterdam is ranked in seventh place from the nine cities studied. The population living in the coastal area is smaller and has a high social resilience. Marseille and Casablanca are less socially vulnerable to coastal floods.

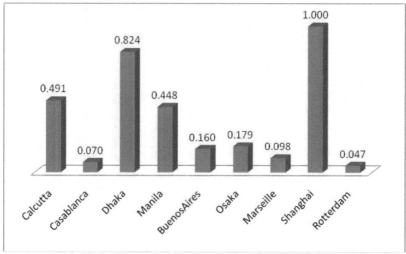

Figure 6.4 CCFVI Social Component

6.5.4. The Economic Component

Three indicators are used to determine the economic CCFVI values. They are: growing coastal population, recovery time after flood event, kilometers of drainage canal. The results of the economic component are shown in Figure 6.5. It can be seen using these FVI criteria that Manila is the most vulnerable economically to coastal floods. This is mainly due to the high number of days needed to recover after a flood event and small kms of drainage, the economy of Manila will recover very slow. The second FVI most economic vulnerable city to coastal floods is Calcutta for similar reasons. Dhaka is the third most vulnerable, Shanghai is situated the fourth place, while Casablanca and Buenos Aires come next. Marseille is the seventh vulnerable coastal city from the nine. Rotterdam and Osaka are least vulnerable from the used indicators point of view. Their economy will recover fast, due to large amount of investment in counter measures and high GDP/capita, the cities show a small exposure to natural hazards, but they have large number of kilometerss of drainage.

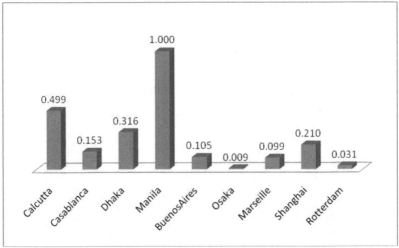

Figure 6.5 CCFVI Economic Component

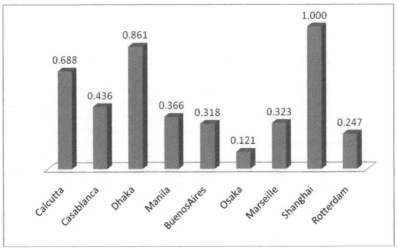

Figure 6.6 CCFVI Politico-Administrative Component

6.5.5. The Politico-administrative Component

The politico-administrative component (PAC) shows the involvement of institutional organisations in the flood management process. As seen in Figure 6.6, the most FVI vulnerable politico-administrative is Shanghai, having small politico-administrative resilience (0.15), a small number of institutional organisations as well as being highly exposed to flood hazards, the uncontrolled planning zone indicator is high 0.6% compared with the other cases where it oscillates around 0.2. Dhaka and Calcutta rank as having the second and third politico-administrative coastal flood vulnerability, 0.25 respective 0.29 resilience value with smaller number of institutional organisations and little flood protection. The cities of Marseille, Rotterdam and Osaka lie in developed countries (UN classification) and have the lowest administrative vulnerability to coastal floods, these cities are the least exposed from PAC

6.5.6. The Overall Coastal City Flood Vulnerability Index

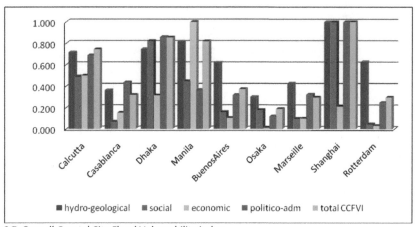

Figure 6.7. Overall Coastal City Flood Vulnerability index

Of the nine cities examined, Shanghai, in China, from the FVI point of view, is most vulnerable to coastal floods overall. Exposed to hydro-geological factors such as storm surge and sea level rise, the city faces a high river discharge, and serious land subsidence, (it results in lowering the standard of coastal flood preventing establishments and increases the risk of natural disasters of typhoon, rainstorm, flood). The indicator soil subsidence is considered during the whole study as an indicator that belongs to the hydro-geological component. From the overall results can be seen that land subsidence indicator influences among other the ranking. From a social perspective, the population density close to the coastline is high, usually experienced floods, but the city does not have high resilience, number of shelters is low compared to the population density. Dhaka, which sits just meters above current sea levels, is regularly impacted by tropical cyclones and flooding, and has very little resilience. Manila in the Philippines and Calcutta in India, are also highly vulnerable cities and tie for the third rank, largely because of the size of the cities, degree of exposure (both experience frequent flooding), and relatively low resilience. Buenos Aires and Casablanca are fourth/fifth, largely because Casablanca is economically vulnerable to floods and has very little flood protection, while Buenos Aires has very low resilience and fewer institutional organisations. Marseille in France and Rotterdam in the Netherlands have equally low vulnerability, mostly because both have slightly more resilience than the other cities, even though the hydro-geological indicators are still significant. Osaka by using the CCFVI criteria is the least vulnerable city out of nine, the city is the least vulnerable hydro-geological and politico-administrative terms.

The advantage of the index is that one can clearly compare vulnerabilities of cities. Poor cities may wish to compare their position relative to rich cities. The CCFVI can be used as a network of knowledge to learn from each other and to increase the resilience of delta cities worldwide through the knowledge given by delta alliances. Publishing the component parts of the coastal city flood vulnerability index can show where progress needs to be prioritised.

The CCFVI model can give a simplified way of characterising what in reality is a very complex system. Such results will help to give an indication of whether a system is resilient, susceptible or exposed to flooding risks and help identify which measures would reap the best return on investment under a changing climate and population and development expansion. The important point is that such a model is used as one tool among others within the whole process of deciding on a roadmap for flood assessment.

On the other hand the method has weakenesses, such as: the index-based technique presents vulnerability over a short period of time. Currently the FVI method is not capable of encapsulating temporal changes; the CCFVI model is a simplification of reality and its application should be compensated with thorough knowledge and expertise-based analysis. The difficulties that the quantification of social and political-administrative indicators, as well as availability of other indicators poses to the calculation may constitute a considerable weakness of the model.

Another potential weakness is the soundness of it underlying assumptions. For the CCFVI to be accepted by a wide range of stakeholders the underlying assumptions for each indicator would need to be explained. The danger in using these types of indices is that stakeholders feel a loss of control and may feel it is just a black box churning out results. A way to overcome this is to involve stakeholders in the developments and in the weighting of the indicators.

It is also a simplified version of reality without capturing the interconnectedness of several indicators and potentially ignoring important local specificities.

6.6. Climate Change impacts on CCFVI indicators and consequently on CCFVI

An indicator-based methodology such as the one outlined here can be used to study the impact of future changes on vulnerability. The indicators express the natural resources, the people and the economic state of the city. Therefore is very important to know the impacts on the people, cities, natural resources, via the use of these indicators.
In particular, it can be used to study the impact of climate change and this is presented below. In order to do this a number of assumptions about the impact of climate change on the relevant indicators has to be made.

The following hydro-geological indicators reflecting climate change projections were chosen: sea level rise (Gornitz, 2001), increasing number of cyclones, higher river discharges and increased storm surge and soil subsidence. The impact of climate change to the social component is reflected in the following indicators: population close to coastline and awareness/preparedness. The indicator "Population close to coastline (UNFPA, 2011a)" will be affected by climate change owing to high population areas which are presently concentrated near the coastline (OECD, 2007; Jun Jian, 2008); if population growth will increase the evacuation of vulnerable populations living in these high risk areas during coastal floods will pose serious problems.

Climate change will have an impact on the awareness/preparedness indicator. UNEP, 2006, presents the importance of communicating with the general public and engaging stakeholders about climate change. Already, educational and public awareness programmes on climate change were developed and implemented, (Russia, Kenya, Albania, Cambodia, etc., UNEP, 2006).

Public awareness aims at early results and is often pursued via the media and outreach campaigns. It is also pursued via education at a more profound, long-term change in habits, particularly among the young. NGOs and journalists can be helpful allies in promoting climate change awareness because of their role as intermediaries with their own widespread networks for outreach. Climate change should worry everybody, but in truth some people will feel more concerned than others because they face particular risks (coastal flooding).

In the present study we assume that the economic component will be impacted by climate change only by one indicator, *growing coastal population*.

Rapid growing population is occurring in coastal cities all over the world. Severe flooding regularly destroys coastal regions, particularly when storm surges and high river flows occur simultaneously. Large coastal cities are particularly at risk from rising sea levels, storms and storm surges, and other aspects of climate change (Fuchs, 2010, Glade, 2003). The densely populated deltas and other low-lying coastal cities are recognized in the IPCC Fourth Assessment Report, 2007 as "key societal hotspots of coastal vulnerability" with lots of people potentially affected.

Only recently local governments and the international development community have seriously begun to consider the implications of climate change on rapidly growing coastal populations and infrastructure. UNEP, 2006 and IPCC, 2007 and their partners initiated programmes to assist coordinated action among scientists, policymakers and the public to support impact and vulnerability assessments, awareness raising about climate change risks and integration of scientific information about impacts, vulnerabilities and adaptation into planning and policy for the affected areas.

It is assumed that the politico-administrative component will not change under climate change impacts. As seen, the political-administrative component comprises four indicators: Flood Hazard Maps, Institutional Organisations, Uncontrolled Planning Zone and Flood Protection. Each of these indicators expresses somehow the political-administrative situation of a coastal area. None of them refer directly to politicians, but to their decisions on investments, etc. Based to the lack of quantification of the relevant indicators by 2100, this component was not investigated in this study.

Scenarios of sea level rise, now – 2100 (UNEP, 1995, IPCC, 2007)

Based on the ranges in the estimate of climate sensitivity and ice melt parameters, and the full set of IS92 emission scenarios, the models predict an increase in global mean sea level of between 13 and 94cm, also an increase in soil subsidence between 1400 and 34000 km2 (horizontal) for the selected case studies. Webster et al, 2005, IPCC, 2007, evaluated the increased number of cyclones and increase of storm surge, using numerical simulation and predicted that by 2100 the number and the intensity of cyclones would increase between 10 and 20%. River discharges will also increase (Jun Jian, 2008) and here it is assumed, based on the IPCC report (IPCC, 2007), that the river discharge will double for all case studies in the worst case scenario.

Two scenarios were used for the hydro-geological component (according to Figure 6.3). The first scenario, is termed the "Best Case Scenario" and assumes that the indicators impacted by climate change increase by 20 to 50% dependent on city location and taking into consideration the values assumed in the literature. In the second scenario, termed the "Worst Case Scenario", the indicators are assumed to have the highest values found in the literature for each case study. For some cities such as Casablanca and Marseille they were assumed to be doubled.

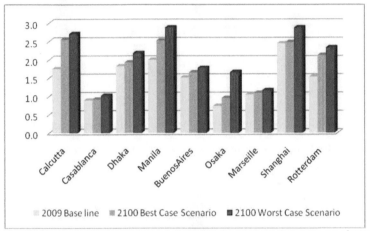

Figure 6.8 Climate change impacts scenarios on Hydro-geological component of CCFVI

By the 2100s, the most vulnerable from the nine cities in terms of hydro-geological exposure can be seen from Figure 6.8.

Again, for the social and economic component, we adopt two scenarios "Best Case Scenario and Worst Case Scenario". For the best case scenario is supposed that the population close to coastal line (social)/growing coastal population (economic) is increasing from 1.1 to 2, depending of values found for those indicators in the Jun Jian (2008)/SRES scenarios for the studied cases. For the worst case scenario we assume that the population close to coastal line (social)/growing coastal population (economic) will triple based on figures from the OECD (OECD, 2007). The results can be seen in Figures 6.9 and 6.10.

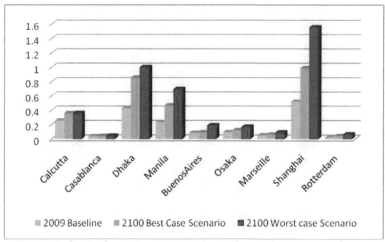

Figure 6.9 Climate change impacts scenarios on the social component of CCFVI

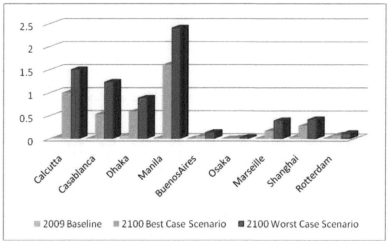

Figure 6.10 Climate change impacts scenarios on the economic component of CCFVI

It can be noted that the component of vulnerability most impacted by climate change is the economic component with high increases in the 2100 results: 6.5 to 177 times the 2009 values. However due to low 2009 values for economic component, the overall 2100 raw results will be influenced in the same way for all the components (see Figure 6.10).

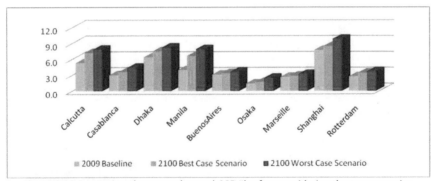

Figure 6.11 Comparison between the total CCFVI's after considering the two scenarios

By 2100, it can be seen in Figure 6.11, as the FVI indicates that the city of Shanghai and Dhaka will remain the most vulnerable to coastal floods, followed tied by Manila and Calcutta, Casablanca, Rotterdam. Buenos Aires and Marseille remain in the lower positions of the nine. Osaka is the city least vulnerable to floods from the selected case studies, very high protection level, high number of kms of dikes.

Also by 2100, it can be seen that the vulnerability of Manila will increase by 2.0, followed by Osaka (1.7), Casablanca (1.5), Rotterdam and Calcutta (1.4), the cities of Dhaka and Marseille will increase by 1.3, Shanghai (1.27) and Buenos Aires (1.2) will increase the least from the nine cities studied. The numbers represent the ratio between Worst Case Scenario and baseline. The position of the city in the rank will not change directly with the increase.

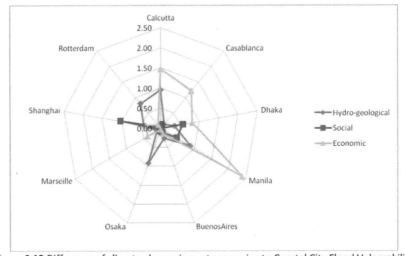

Figure 6.12 Difference of climate change impacts scenarios to Coastal City Flood Vulnerability

The baseline 2009 results show the cities' vulnerability from 2009. In order to see which component is most vulnerable to flooding for each coastal city with the 2100 worst case scenario the values are shown without normalisation. The difference between the raw CCFVI results of the two cases were computed and the variations are presented in Figure 6.12. The FVI shows which cities are most impacted with climate change in relation to each component. Manila has the most variations which reflects the fact that the city is characterised by economic extremes with a considerable disparity between the wealthy few and the large number of people below the poverty line. Reportedly, 97% of the total GDP in the Philippines is controlled by 15% of the population (Yusuf and Francisco, 2009). With the Growing Coastal Population indicator increasing by 2100 Manila city will be even more economically vulnerable to coastal floods, which explains the high economic vulnerability.

Table 6.5 Total FVI, ranking of coastal cities for different scenarios

Coastal Cities Total FVI Ranking			
City	2009 Baseline	2100 Best Scenario	2100 Worst Scenario
Calcutta	3	3	4
Casablanca	6	5	5
Dhaka	2	2	2
Manila	4	4	3
BuenosAires	5	7	7
Osaka	9	9	9
Marseille	8	8	8
Shanghai	1	1	1
Rotterdam	7	6	6

The study also provides interesting insights into future (2100) total flood vulnerability on a coastal city scale (Total CCFVI), a ranking of those cities can be see in Table 6.5. The FVI indicates that still by 2100 worst case scenario the most vulnerable city to coastal floods is Shanghai, followed very close by Dhaka and least vulnerable Marseille and Osaka, taking into account that no adaptation measures are included into the analysis.

6.7. Discussion on managing coastal cities throughout CCFVI tool (http://unesco-ihe-fvi.org)

There can be no analysis or assessment of flood and risk management and adaptation measures without, first, understanding correctly the concept of vulnerability. Vulnerability is an analysing element of exposure, susceptibility and of resilience of any coastal system at hazard. The vulnerability index captures a zooming in view, which helps in assessing aspects that might have been neglected by traditional approach to assess or analyse risk so far. As another argument to assess and index vulnerability is the requirements of the European Floods Directive 2007/60/EC and the 'Hyogo Framework for Action 2005–2015. Therefore, not only politicians, but also decision makers need such indices, the standardised vulnerability indices are helping in assessing and monitoring the "elements at risk" (Merz et al., 2007). This need, the indices, involves a range of subjective decisions, the choice of indicators for example, but if the vulnerability concept is well defined and clarified the evaluation and interpretation of composite indices will bring us the understanding of where to mitigate risk and where to focus investments. Still there is no standardised way to measure vulnerability (Bohle et al. 1994), the measurement depends of each hazard that occurs, its frequency and its intensity.

The methodology used in this chapter is based on sets of indicators for different vulnerability factors and coastal system components, focusing on coastal floods. Selected indicators (19) have been used to evaluate coastal cities flood vulnerability. In the CCFVI approach the indicators were normalised separately using predefined minimum and maximum values to facilitate easily comparison between case studies and to provide one single and comparable value. The final Coastal City Flood Vulnerability component values are therefore between 0 and 1. However in order to present consistent final results of climate change impact scenarios the raw results are shown.

Since the methodology is based on indicators, its main weakness is the accuracy of the data on which it is based. For the results to be valid, all data must be derived from reliable sources (see Section 6.5.1). This approach allows for relative comparisons to be made between urban areas irrespective of uncertainties. In this way proposed measures can be prioritised for urban areas that are at greatest risk. Uncertainty is not removed, but is integrated into the assessment. It also offers a more transparent means of prioritising, which is inevitably a highly political process.

It is worth underlining that the CCFVI is a flexible tool: it can be used to create different 'scenarios' by changing one or more indicators and can be tailored on different situations and areas, since the principle 'one size fits all' cannot be applied to vulnerabilities present in complex and dynamic realities.

Another weakness of the method, the unevenness of the indicators has two main consequences. Firstly, the indices calculated by the model are distorted. Components with more indicators are more reliable than components with few indicators on this level. Components with few or no indicators are not well-represented. Furthermore, the components with few indicators are less reliable. This is due to the occurrence that a large variation in one variable has a larger influence on the calculated index. The second main consequence of the unequal distribution of indicators is on the feasibility of calculating the index of a certain component. If more indicators are available, the user of the model can choose the indicators according to the information that is most readily available making the model more feasible.

To improve the model, the number of indicators would need to be increased in the areas with fewer indicators at the moment. Preferably, these indicators would coincide with information which most governing bodies are already collecting in order to increase the ease of use of the model.

The CCFVI can be used as an effective tool to assist decision-makers in evaluating the impacts of different scenarios. An evaluation of coastal flood vulnerability is the CCFVI, a non-structural measure, (more in the interest of general public, policy and decision makers, re/insurance companies), the index can be incorporated into the Disaster Risk Reduction Strategy.

The CCFVI can also be used as a toolkit to assess and manage the coastal flood vulnerability and this way to facilitate adaptation and coping capacities. The CCFVI methodology is at the base of a network of knowledge, among the flood vulnerability index methodology (unesco-ihe-fvi.org), which is an automated calculation of a CCFVI implemented through a web management interface (PHP) that enhances the ability of decision makers to strategically guide investment. The network of knowledge can be used between different institutions, universities and non-governamental organizations with the purpose to encourage collaboration between the members of the network on managing coastal flood vulnerability information and also promoting further studies on flood risk assessment at smaller scales.

The CCFVI can help in converting knowledge into actions: to assess/index the coastal flood vulnerability in different regions of the world, but also in less well served ones, data scarce areas, by following an integrated approach; raise coastal flood vulnerability awareness; save lives, reduce economic, environmental losses and distribute the financial burden better.

Other future developments work two ways. Firstly, increasing populations are likely to increase the vulnerability to flooding owing to increased inhabitation of flood-prone areas (Saalmueller, n.d.). Populations are likely to expand into areas which are currently covered by natural vegetation (Marcoux, 2000; DeFries et al. 2010) and the densities of urban environments already within flood-plains and coastal areas are likely to increase (UNFPA, 2011b).

On the other hand, the vulnerability could decrease due to development which could lower susceptibility indicators such as increased awareness and education (IIASA, 2008). Other resilience indicators could also improve due to public investments in (protective) infrastructure and increased institutional capacity as countries develop.

6.8. Conclusions

The conclusions of this chapter covers three aspects:

- CCFVI methodology and use: The use of the CCFVI can make inhabitants and governments aware of vulnerability in their area. This way it helps policy makers and water authorities to define what measurements should be taken. CCFVI is a powerful tool for mapping of vulnerable areas within the cities. Thus the correct use of CCFVI can help policy makers and urban planners in making decisions with regards to development in specific areas and possible funding allocation for adaptation and reduction of flood vulnerability in urban areas. The continuous monitoring of the CCFVI for particular urban areas may also show a trend in the development of the area over time and will also give tangible information for preparation in case of flooding. Thus, the CCFVI is necessary, but not sufficient for decision making. The CCFVI has to be used in combination with other decision-making tools, which include participatory methods with the population of areas identified as vulnerable and expert judgment.

- CCFVI baseline results: in comparing the cities, Shanghai came out as the most vulnerable to coastal floods. Rotterdam (the Netherlands) and Osaka (Japan) were the least vulnerable to floods. The poorest cities have a very low resilience to floods, are the most exposed socially and have weak institutional organisations

- Climate change impacts on the CCFVI: climate change is expected to impact the urban coastal delta in various and diverse ways. The *hydro-geological component* is clearly one of these. However, other components are also affected. The *social component* shows that the social vulnerability to floods of Shanghai, Dhaka and Manila will double by 2100. The population close to the coastline and the amount of cultural heritage exposed to floods will increase, so there is an urgent need for action towards adaptation measures by raising the awareness of local population. The third component affected is the economic component which is more sensitive. By 2100 the economic vulnerability of the delta cities will certainly increase. Actions must be undertaken to protect them in order to reduce their vulnerability to floods. Collaboration between delta cities' administrations, multiple stakeholders and organisations at international level (delta-alliances) have to be undertaken to support the most vulnerable areas and to learn from each other. The use of CCFVI and climate change scenarios offer the opportunity to get a broad overview on components affected and on possible adaptation options that could be applied, directing resources at more in-depth investigation of the most promising adaptation strategies. At a later stage, it can also serve to evaluate the effectiveness of adaptation measures.

CHAPTER 7

DISCUSSIONS AND CONCLUSIONS

7.1 Summary of objectives

This chapter discusses the overall achievements of the presented research in terms of the general and specific objectives as well as the contribution to science and society.

The objective for this thesis was formulated as:

> *To demonstrate the applicability of the improved FVI methodology, so it can contribute to the development of the existing knowledge base on flood risk assessment methodologies.*

Five specific objectives were formulated:

1. To develop and apply the FVI methodology to various spatial scales, such as: river basin, sub-catchment and urban area.
2. To analyse and reduce the complexity of the FVI, through various mathematical and statistical methods.
3. To develop and apply the FVI for coastal cities based on existing approach.
4. To investigate the possibility of comparing the FVI methodology versus traditional flood modelling techniques.
5. To gain academic acceptance by creating a network of knowledge in different countries with different socio-economic situations, based on the FVI.

Through the development of the river flood vulnerability index/coastal flood vulnerability index this thesis hopes to contribute to identify and develop action plans to deal with floods and flooding or on smaller scales to improve local decision-making processes by selecting measures to reduce vulnerability at local and regional levels. In this way, the need to contribute, to the protection of the river/coastal system and of the social-economic component is depending on this system.

The following sections first discuss the specific objectives and their conclusions, and subsequently discuss the contribution to science and society. The chapter concludes with an overall conclusion for this thesis and future works.

7.1.1 Specific Objective 1 and 5: the development, implementation and dissemination the FVI for diverse spatial scales

The conclusions concerning the development of the FVI methodology can be summarised as follows:

- By defining and evaluating flood vulnerability, it has become a more tangible concept. It has become clearer what the factors of flood vulnerability are. This knowledge enabled the evaluation of the FVI for flood risk management and of the advantages and disadvantages of FVI methodology.

- Applying FVI in flood risk management entails adopting a systems approach in which the reaction of the river basin, sub-catchment and urban area together to the whole FVI components is considered by calculating the values of the exposure, susceptibility and resilience indicators. This 'FVI approach' includes a thorough analysis of water resource system and its relation with floods, especially flood vulnerability.

- The proposed methodology to calculate a FVI provides an approach to assess how much floods affect, or can affect natural or build systems.

- The use of the FVI methodology improves the decision-making process by identifying the vulnerability of flood prone areas; it is concluded that for the larger scales, trans-boundary river basins, action plans are need to deal with floods and flooding in order to identify and develop; and for the smaller scales is necessary to improve the (local) decision making process by selecting action plans to reduce vulnerability at local and regional scales. It is also conclude that a more in-depth interpretation of local indicators which pinpoints actions is needed to diminish focal spots of flood vulnerability.

- FVI offers easy to comprehend results, with the use of a single value to characterize high or low vulnerability.

- The FVI is a powerful tool for policy and decision makers to prioritise investments and makes the decision-making process more transparent. Identifying areas with high flood vulnerability may guide the decision-making process towards better means of dealing with floods.

The network of knowledge – implementation and dissemination of FVI methodology

The FVI and CCFVI methodologies used on the three spatial scales for river and coastal floods, have been disseminated via the FVI website (http://unesco-ihe-fvi.org, Chapter 3), and the results serve as references to enable users to apply it in different parts of the world. Through this website, the FVI/CCFVI can help to assess and improve the links of the safety chain in risk management.

In conclusion:
- It is believed that the FVI website has been very useful in developing the index further. Over time, different methodologies can be developed and incorporated into the website. The tool will present a means of assessing vulnerability in a future that is uncertain.

- The network of knowledge can be used between different institutions, universities and non-governmental organizations with the purpose to encourage collaboration between the members of the network on managing river/coastal flood vulnerability information and also promoting further studies on flood risk assessment at smaller scales.

- As the FVI, also the CCFVI can to be used as a toolkit to assess and manage coastal flood vulnerability and to facilitate adaptation and coping capacities. The CCFVI can help in converting knowledge into actions. Such us: assess/index the coastal flood vulnerability in different regions of the world, but also in less well served ones, data scarce areas, by following an integrated approach; raise coastal flood vulnerability awareness; save lives, reduce economic, environmental losses and distribute the financial burden better.

- The FVI is necessary, but not sufficient, for decision making and therefore should be used in combination with other decision-making tools. This should specifically include participatory methods with the population of areas identified as vulnerable and should also include a team of multidisciplinary thematic specialists and knowledgeable societal representatives and those with expert judgment.

7.1.2 Specific Objective 2: the reduction of the FVI complexity

The flood vulnerability index methodology developed by in this thesis uses 71 indicators in its calculation. However, it is recognized that some of these indicators may be redundant or have no influence on the results. This specific objective was intended to select the most significant indicators in order to establish parsimonious usage of the FVI (Chapter 4).

The major conclusions for a good practical assessment can be summarised as follows:

- By analysing the existing indicators for each spatial scale and for each vulnerability component, only the most significant ones were retained. A reduced number, using mathematical methods, of indicators is needed to sort out the essential key indicators for a simpler, easier and low-cost application. Analysing the indicator's significance through a survey was also carried out to portray reality in an effective way. The survey results were combined with the mathematical ones to obtain appropriate FVI indicators in order to easily formulate the FVI equations.

- After combining these methods, it was noticeable that the environmental component of the river basin and sub-catchment is more realistic as the flood vulnerability ranking changes for the case studies using the new equation. This change is in a better way for some case studies and worse for other case studies. Once updated, the environmental set of indicators gives a clearer overview of the environmental FVI on water conditions as well as land coverage.

- Simplifying the FVI makes it more usable and gives users more confidence that it is giving the correct trends; can be used as a tool for decision making to direct investments to the most appropriate sectors and also to help in the decision-making process relating to flood defence, policies, measures and activities.

- Any natural system has its own uncertainty. While a level of uncertainty is inherent in FVI, the use of it in operational flood management is highly relevant for policy and decision makers in terms of starting adaptation plans. It offers a more transparent means of making such priorities, which inevitably are considered as highly political decisions. It may also be considered as a means to steer flood management policy in a more sustainable direction. However, as individual information is lost in the aggregation process, it needs to be retrieved by a more in-depth analysis of each process in order to design policies and their implementation.

7.1.3 Specific Objective 3: the development and application of the CCFVI

This specific objective refers to the development of Coastal City Flood Vulnerability Index (CCFVI) based on exposure, susceptibility and resilience to coastal flooding (Chapter 6). It is applied to nine cities around the world, each with different kinds of exposure. This specific objective as well refers to the impact of climate change on the vulnerability of these cities over a longer timescale.

The conclusions of this work covers three aspects:
- The use of the CCFVI can make inhabitants and governments aware of vulnerability in their area. With the CCFVI the impacts can be predicted in different more scenarios. In this way it helps policy makers and water authorities to define what measurements should be taken. The continuous monitoring of the CCFVI for particular urban areas may also show a trend in the development of the area over time and will also give tangible information for preparation in case of flooding. The CCFVI can be used on a smaller geographical unit, to discuss the most relevant factors explaining exposure, and especially susceptibility and resilience.

- The poorest cities have a very low resilience to floods, are the most exposed socially and have weak institutional organisations

- Climate change is expected to impact the urban coastal deltas in various and diverse ways. The *hydro-geological component* is clearly one of these by defintion. However, other components are also affected. The *social component* indicates that the social vulnerability to floods of Shanghai, Dhaka and Manila will double by 2100. The population close to the coast line and the number of cultural heritages exposed to floods will increase, so there is an urgent need for action on adaptation measures by raising the anticipatory mentality of local population. The third component affected is the *economic component* which is more sensitive. By 2100 the economic vulnerability of the delta cities will increase drastically. Actions must be undertaken to protect them in order to reduce their vulnerability to floods. Collaboration between delta cities' administration, multiple stakeholders and organisations at international level (delta-alliances) have to be undertaken to support the most vulnerable areas and to learn from each other. The use of CCFVI and climate change scenarios offer the opportunity to get a broad overview of components affected and possible adaptation options that could be applied, directing resources at extensive investigation of the most promising adaptation strategies. At a later stage, it can also serve to evaluate the effectiveness of adaptation measures.

7.1.4 Specific Objective 4: the comparison of the FVI versus deterministic approach

T his specific objective focuses on the applicability and performance of the flood vulnerability index; the comparison of FVI methodology to the deterministic approach, as seen in Chapter 5. The comparison is undertaken in data scarce area (Budalangi area, Kenya, using the SOBEK 1D/2D model). The FVI method might be particular useful in data scare areas. Examining the results from this lead to the following conclusions:

- It is obviously that the FVI is not assessing directly flood risk, but has a contribution in evaluating the risk; risk covers only the economic consequences whilst vulnerability takes a step further and covers some other aspects, such as: social, environmental and physical.
- The parametric approach, here the FVI, through indicators is the only one which assesses vulnerability to floods; the deterministic approach has a better physical basis, but limited evaluation of vulnerability; FVI gives a wider evaluation, but is less physically rigorous. Therefore FVI is useful in a larger-scale vulnerability assessment, whilst a deterministic approach is better for more focused studies. In fact FVI could be used to decide where a deterministic model is necessary.
- The Flood Vulnerability Index as analysed in the research provides a quick and reliable method to assess vulnerability to floods; it is the only method for assessing the vulnerability to flooding of a particular geographical area. The fact that indicators are calculated and used, allows for comparison of flood vulnerability in different areas as well as the identification of which indicators can determine the relative level of flood vulnerability. FVI can measure trends in the changing natural and human environments, helping identify and monitor priorities for action. These features, alongside the ability to identify the root causes of increased vulnerability, provide key information at a strategic level for flood risk planning and management. However the results would provide neither sufficient information nor the required level of detail for input into engineering designs or project level decisions.

The complex developments and dynamics in systems are not easy to include in the models.

- FVI can provide an insight as to the most vulnerable locations. It can analyse the complex interrelation among a number of varied indicators and their combined effect in reducing or increasing flood vulnerability at a specified location. It is very useful when there is a large level of uncertainty and decision makers are faced with a wide array of possible actions that could be taken in different scenarios, in this case the FVI can present an easy to understand and communicate results that would assist decision makers in identifying the most corrective/effective measures to be taken. In this way proposed measures can be prioritised for areas that are at greatest risk. Uncertainty is not removed, but is integrated into the assessment. On the other hand this complexity is a negative point as well, since it takes a long time and good knowledge of the area and the system behind the FVI to be able to implement it.
- However, as with all models, the FVI model is a simplification of reality and its application should be compensated for with thorough knowledge and expertise-based analysis. The difficulties that the quantification of social indicators, as well as availability of other indicators poses to the calculation may constitute a considerable weakness of the model. With all of these, the FVI assesses the social and environmental vulnerability of a system.
- FVI is a planning tool for risk assessment - the FVI represents a probability. At the same time it can give a false impression of certainty and it can be questioned whether a small set of indices for any river system for example, really contains any valuable information whatsoever.
- Another potential weakness of this parametric approach is the soundness of the underlying assumptions: for the FVI to be accepted by a wide range of stakeholders it underlying assumptions for each indicator would need to be explained. The danger in using these types of

indices is that stakeholders feel a loss of control and may feel it is just a black box churning out results. A way to overcome this is of course to involve stakeholders in the developments of the indicators.

- Obviously such a parametric model is limited by the accuracy and availability of good datasets. A number of the indicators are very hard to quantify especially when it comes to the social indicators. On the other hand, such a model can give a simplified way of characterising what in reality is a very complex system, but seems to be the only one which assesses vulnerability. Such results will help to give an indication of whether a system is resilient, susceptible or exposed to flooding risks and help identify which measures would reap the best return on investment under a changing climate and population and development expansion. The important point is that such a model is used as one tool among others within the whole process of deciding on a roadmap for flood assessment.

7.2 Contribution to science and society

The five specific objectives have been discussed, but what has this thesis contributed to science and society? From Chapter 1 it has become clear that with regard *to science*, the study should contribute to flood risk management, flood mitigation and decision making processes. With regard *to society*, indirectly, flood mitigation and decision making processes will help the exposed, susceptible and less resilient communities to deal with river and coastal floods. The contribution of this thesis to these issues is discussed in this section.

7.2.1 Contribution to flood risk management, flood mitigation and decision making processes

The main contribution is that the parametric approach, the FVI, through indicators is the only way to assess vulnerability to floods; this in the context when evaluating vulnerability is a requirement of the European Floods Directive 2007/60/EC. The flood risk management strategies should "focus on prevention, protection and preparedness" it is one of the aims to reduce risk from natural hazards. As the European Floods Directive, in Kyoto, 2005, the 'Hyogo Framework for Action 2005–2015: Building the resilience of Disaster Reduction (WCDR) requires the need to "identify, assess and monitor disaster risks' (UN 2005: 12). To achieve this goal, the Kyoto declaration stressed the development of an indicator based systems of disaster risk and vulnerability for multiple scales. Therefore the indicators, all through an index, can be a guide to understanding in a holistic way the current state of a system, also indicating the possible strategies to improve the functioning of the system.

The method developed in this thesis provides a structured approach to carry out an assessment of vulnerability, but "not a quantification of vulnerability" especially flood vulnerability related to water resources system, via flood vulnerability index, which is applicable for river floods in three different spatial scales (river basin, sub-catchment, urban area) and for coastal floods at urban area scale. Throughout this FVI, individually, the four components of the water resources system can be assessed, even more for each system can be known the exposure, susceptibility and resilience.

Evaluating the four aspects of a system, regarding vulnerability to floods by using an easy to apply, but complex approach, based on indicators, is considered to be a contribution to the flood risk management, flood mitigation and a transparent way to guide decision-makers. Flood vulnerability assessment is important to give systems and related communities warnings about their vulnerable position. This approach can support researchers and consultants in the field of flood risk management and flood mitigation. This flood vulnerability approach includes a thorough analysis of society/system and its relation with floods. Subsequently, this may enhance informed decision-making with regard to water resources management.

The socio-economic component is highly complex. This thesis has tried to provide a view in the contribution of flood vulnerability index to societies, and in this way hopes to contribute to the flood protection of the local communities. Yet, due to the complexity and dynamic character of the socio-economic component of the system, there is a risk that not all impacts are considered or are not quantified to the correct extent. Careful monitoring in combination with adaptive management plans is therefore pertinent.

It is important that decision makers are aware of the importance of FVI/CCFVI for water resources systems and societies; and of the methods to assess flood vulnerability and the impacts on societies. This thesis is meant to provide a holistic approach to be used in such an assessment, and this way hopes to facilitate the consideration of system impacts in water resources decision-making.

7.3 Final conclusion

The thesis intended to provide an approach for assessing flood vulnerability indices of a river and coastal system as part of flood risk management, consisting of a conceptual methodology together with a stepwise and holistic approach (identifying indicators, factors of vulnerability, components of the system, application to different spatial scales), as well as a tool and indirect guidance for the flood vulnerability assessment. With this approach the thesis makes a contribution to the further optimisation and management process of flood vulnerability, flood mitigation and implicitly flood risk management. The approach presented in this thesis generates more comprehensive, transparent and more socially-relevant information to decision-makers.

To conclude, the FVI developed shows that the FVI tool can be applied in a broad range of contexts (river and coastal floods, including their conditions of the applied index components and indicators), and can produce helpful understanding into vulnerability and capacities for using it in planning and implementing projects. The FVI presents vulnerability only in a short window in time space. Because some of the shown data cannot be the most recent, for example, climate change processes. The FVI was developed to prioritise investments and to respond to a flood disaster by understanding what impact interventions will have on vulnerabilities in place. Being a parametric model, the FVI gives tangible results in assessing flood vulnerability at each scale. It is definitely worth noting that the latest data available and applied actively helps to encapsulate local and temporal elements.

It is clear that the FVI is not assessing directly flood risk, but has a contribution in assessing the risk; flood risk relates to "human health, the environment, cultural heritage and economic activity" (Scottish Government, 2009) since vulnerability takes a step further and covers some other aspects, such as:
- social (relates to two factors: on the one hand the presence of human beings which encompasses issues related to, for example, deficiencies in mobility of human beings

associated with gender, age, or disabilities; on the other hand floods can destroy houses, disrupt communication networks, or even kill people. Included in this component are the administrative arrangements of the society, consisting of institutions, organizations and authorities at their respective level),

- environmental (deforestation, urbanization and industrialization have enhanced environmental degradation) and
- physical (relates to the predisposition of infrastructure to be damaged by a flooding event).

The parametric approach, here the FVI, through indicators is the only one which evaluates vulnerability to floods; any other deterministic approach has a better physical basis, but limited evaluation of vulnerability; FVI gives a wider evaluation, but is less rigorous. Therefore FVI is useful in a larger-scale vulnerability assessment, but a deterministic approach is better for more focused studies. In fact FVI could be used to decide where a deterministic model is necessary.

The FVI provided new knowledge to understand, apply, asses, manage and mitigate vulnerability, firstly to floods and lately to other natural disasters.

7.4 Future research

This thesis describes a flood vulnerability index, indicators based methodology, by identifying proper indicators can help in assessing the flood vulnerability of a defined spatial scale. The research for this thesis has been done from flood mitigation and flood risk assessment point of view. Now that concepts of vulnerability factors, system's components and spatial scales are better understood, it would be interesting to investigate to what extent this approach can be used in an opposite natural disaster, such as droughts. A drought vulnerability index can be developed, and together with the FVI to produce combined vulnerability maps.

Alternatively, novel ways have been derived and tested for measuring flood vulnerability to individual, by approaching it in an inter-disciplinary manner, bringing together engineering and social approaches. Each human being has their way of thinking regarding floods, i.e. individuals of the same age and education can have a different approach and speed to recovering from floods, and different societies approach recovery differently. The method could be developed using a vulnerability-based approach driven by three main integrating factors: i) individual exposure, the magnitude and frequency of flood considered, ii) susceptibility to floods/their vulnerability perception, iii) resilience. A novel toolkit could be developed to identify the cognitive actions and preparedness of individuals to floods based on the vulnerability factors. The methodology could be developed for all types of floods.

The method described in this thesis has been applied in many case studies. However, the results of these case studies have not been used in real decision-making processes yet. It will be useful to learn to what extent the FVI generates information which can actually help to better-inform decision-making and whether different decisions are or would be taken into consideration as a results of FVI.

As a future research, the FVI could be integrated into the safety chain approach in flood risk management policy. The safety chain distinguishes between five links or phases of risk and crisis management: pro-action, prevention, preparation, response and recovery. The FVI can be seen as an important tool in the context of SEA (Strategic Environmental Assessment) to communicate impacts and vulnerability and to evaluate development alternatives to adapt to the changes. It would be useful to show importance of FVI within the future flood risks, use the FVI to prevent the further harm of the water resources system and to avoid or reduce costs and efforts from the flood risk management.

REFERENCES

A Commonwealth vulnerability index for developing countries: the position of small states By Jonathan P. Atkins, Sonia Mazzi, Christopher D. Easter

Abuodha, P and Woodroffe, CD. 2007. Assessing vulnerability of coasts to climate change: A review of approaches and their application to the Australian coast, In Woodroffe CD, Bruce, E, Puotinen, M and Furness RA (Eds), GIS for the Coastal Zone: A selection of Papers from CoastGIS 2006, Australian National Centre for Ocean Resources and Security University of Wollongong, Wollongong, Australia, 458.

ADB, 2004, Asian Development Bank Annual Report, 2004, Statutory Reports and Official Records, ISSN: 306-8370

Adger N.W., 2006, Vulnerability, Global Environmental Change 16 pg 268–281

Adger, N., Kelly, M.P., Huu Ninh, N., 2004, Living with Environmental Change: Social Vulnerability, Adaptation and Resilience in Viet Nam, Journal of Comparative Economics 32, pp. 367-369

Adger, N.W., 1999, Social vulnerability to climate change and extremes in coastal Vietnam. World Development 27: 249–269.

Adger, W. N., Brooks, N., Bentham, G., Agnew, M. and Eriksen, S., 2004. New Indicators of Vulnerability and Adaptive Capacity. Tyndall Centre for Climate Change Research,University of East Anglia, Norwich, UK.

Adger, W. N., Hughes, T. P., Folke, C., Carpenter, S. R. and Rockstrom, J., 2005. Social-ecological resilience to coastal disasters. Science, 309. 1036–1039.

Adger, W. N., Vincent, K., 2005, Uncertainty in adaptive capacity, C.R. Geo-science 337, pp. 399-410

Adriaanse, A., 1993. Environmental policy performance indicators: a study on the development of indicators of environmental policy in the Netherlands. SDU Publishers, Amsterdam and The Hague.

Ahrendt, K. 2001. Expected effects of climate change on Sylt Island; results from a multidisciplinary German project. *Climate Res.*, 18, 141-146.

Alexander, D., 1993, Natural disasters, New York, Chapman & Hall.

Aller L, Bennet T, Lehr JH, Petty RJ, Hackett G (1987) DRASTIC: a standardized system for evaluating ground water pollution potential using hydrogeological settings. National Water Well Association, Dublin, Ohio, 266 pp

AMMA, 2009, African Monsoon Multidisciplinary Analyses, Characteristics of the West African Monsoon, as seen on: http://www.amma-international.org/article.php3?id_article=10, on November, 2010

Ancey, C., 2010, Hydraulique a surface libre. Phenomenes de propagation : ondes et ruptures de Barrage. Bases mathematiques, outils de simulations, applications. Notes de cours, version 2.4 du 15 mai 2010, Laboratoire hydraulique environnementale (LHE) Ecole Polytechnique Federale de Lausanne Ecublens CH-1015 Lausanne

Andjelkovic I., 2001, International Hydrological Programme Guidelines on non-structural measures in urban flood management, IHP-V Technical Documents in Hydrology No. 50 as see on http://unesdoc.unesco.org/images/0012/001240/124004e.pdf 8th of October 2009

ASCE & UNESCO, 1998. Sustainability Criteria for Water Resource Systems. ASCE, Reston, VA, USA.

Atkins, J., Mazzi S and Ramlogan, C. (1998), A Study on the Vulnerability of Developing and Island States: A Composite Index Issued by the Commonwealth Secretariat, August 1998.

Atkins, J.P.;Mazzi, S.; and Easter,C. (2001): Small States: A Composite Vulnerability Index. In: Peretz,D.; Faruqi, R.; Eliawoni, J. (Eds.), Small States in the Global Economy.Commonwealth Secretariat Publication,pp.53-92.

Baarse, G. 1995. Development of an Operational Tool for Global Vulnerability Assessment (GVA): Update of the Number of People at Risk due to Sea-Level Rise and Increased Flood Probabilities. CZM-Centre Publication No. 3, Ministry of Transport, Public Works and Water Management, The Hague, The Netherlands.

Balica S.F., Wright N.G., (2010). Reducing the complexity of Flood Vulnerability Index, Environmental Hazard Journal , Volume 9, Number 4, 2010 , pp. 321-339(19

Balica, S.F. 2007. Development and application of flood vulnerability index methodology for various spatial scale, MSc thesis (WSE-HERBD 07-01), UNESCO-IHE, Institute for Water Education, April 2007.

Balica, S.F., Douben, N., Wright, N.G. (2009). Flood Vulnerability Indices at Varying Spatial Scales, Water Science and Technology Journal, vol. 60, no10, pp. 2571-2580, ISSN 0273-1223

Balica, S.F., Wright, N.G. 2009. A network of knowledge on applying an indicator-based methodology for minimizing flood vulnerability, Hydrological Processes. Volume 23 (20), pp. 2983-2986, 2009.

Barnett, J. and Adger,W. N., 2003. Climate dangers and atoll countries. Climatic Change, 61. 321–337

Barredo, J.I., de Roo, A., Lavalle, C., 2007, Flood risk mapping at European scale. Water   Science   and Technology, Vol. 56, 4, 11-17

Barroca, B., Bernardara, P.,Mouchel, J.M. and Hubert, J. M., 2006. Indicators for identification of urban flooding vulnerability. Natural Hazards and Earth System Sciences, 6. 553–561. www.nat-hazards-earth-syst-sci. net/6/553/2006 (accessed 3 December 2009).

Bascom, W.H., 1954, Characteristics of natural beaches. Proceedings of the 4th Conference on Coastal Engineering, pp. 163 180. Victoria, Australia: Institute of Engineers of Australia.

Bates, P., De Roo, A., 2000, A simple raster-based model for flood inundation simulation, Journal of Hydrology, 236, 54-77.

Bates, P., Horrit, M., Hunter, N., 2005, LISFLOOD-FP User manual and technical note Code release 2.6.2, University of Bristol, School of Geographical Sciences, University of Bristol, University Road, Bristol, BS8 1SS, UK. 22nd June 2005

Bates, P.D., Horritt, M.S., Aronica, G., Beven, K., 2004. Bayesian updating of flood inundation likelihoods conditioned on flood extent data. Hydrological Processes 18, 3347–3370.

BBC, 2006, BBC NEWS, Rain deepens Danube flood misery, Thursday 20th April, 2006, as see on http://news.bbc.co.uk/1/hi/world/europe/4927688.stm on 19th of February 2007

Begon, M., Harper, J.L., Townsend, C.R. (1996). Ecology, individuals, populations and communities. Blackwell Science Ltd., Oxford, UK.

Begum, S., Stive, M. J. F. and Hall, J.W., 2007. Flood Risk Management in Europe: Innovation in Policy and Practice. Springer, London, 237.

Bijker, E.W.1996. History and Heritage in coastal engineering in the Netherlands, ed. C. Kraus, New York, 390-412.

Bijlsma, L., C.N. Ehler, R.J.T. Klein, S.M. Kulshrestha, R.F.McLean, N.Mimura, R.J.Nicholls, L.A.Nurse, H. Perez Nieto, E. Z. Stakhiv, R. K. Turner, and R. A. Warrick, Coastal zones and small islands. Climate Change. 1995. Impacts, Adaptations and Mitigation of Climate Change: Scientific-Technical Analyses, Cambridge University Press, Cambridge, 289-324.

Birkmann, J. (ed.), 2006. Measuring vulnerability to promote disaster-resilient societies: conceptual frameworks and definitions. Measuring Vulnerability to Natural Hazards: Towards Disaster Resilient Societies, J. Birkmann (ed.). United Nations University Press, New York.

Birkmann, J., & Wisner, B., 2006, Measuring the Un-Measurable, The Challenge of Vulnerability UNU – EHS, no 5/2006, Bonn, Germany

Blaikie, P., Cannon, T., Davis, I. andWisner, B., 1994. At Risk: Natural Hazards, People's Vulnerability and Disasters. Routledge, London.

Bohle HG, Downing TE, Watts MJ (1994) Climate change and social vulnerability: toward a sociology and geography of food insecurity. Global Environ Change 4(1):37–48

Boruff, B.J., Emrich, C. and Cutter, S.L. 2005. Erosion hazard vulnerability of US coastal counties. *Journal of Coastal Research*, 21: 932-942.

Bosher, L., Dainty, A., Carrillo, P. and Glass, J. (2007), "Built-in resilience to disasters: a pre-emptive approach", Engineering, Construction and Architectural Management, Vol. 14 No. 5, pp. 434-46.

Bosher, L., Penning-Rowsell, E. and Tapsell, S., 2007. Resource accessibility and vulnerability in Andhra Pradesh: Caste and non-Caste influences. Development and Change, 38(4). 515–640.

Bouma J.J, Francxois D., Troch P., 2005, Risk assessment and water management   Environmental Modelling & Software 20 pg. 141-151

Brenkert, A., and Malone E., 2005. Modeling Vulnerability and Resilience to Climate Change: A Case Study of India and Indian States. Climatic Change, 72(1):57–102, September 2005. URL http://dx.doi.org/10.1007/s10584-005-5930-3.

Briguglio, L. (1992), Preliminary Study on the Construction of an Index for Ranking Countries According to their Economic Vulnerability, UNCTAD/LDC/Misc.4 1992.

Briguglio, L. (1993), The Economic Vulnerabilities of Small Island Developing States Study commissioned by CARICOM for the Regional Technical Meeting of the Global Conference on the Sustainable Development of Small Island Developing States, Port of Spain, Trinidad and Tobago, July 1993.

Briguglio, L. (1995), Small Island States and their Economic Vulnerabilities World Development Vol.23(9), 1615-1632.

Briguglio, L. (1997), Alternative Economic Vulnerability Indices for Developing Countries, Report prepared for the Expert Group on Vulnerability Index, United Nations Department of Economic and Social Affairs-UN(DESA), December 1997.

Briguglio, L., 2003. Methodological and Practical Considerations for Constructing Socio Economic Indicators to Evaluate Disaster Risk. Institute of Environmental Studies, University of Colombia, Manizales, Colombia, Programme on Information and Indicators for Risk Management, IADB-ECLAC-IDEA.

Briguglio, L., 2004, Economic Vulnerability and Resilience: Concepts and Measurements, home.um.edu.mt/islands/brigugliopaper_version3.doc

Brinkman, J. (2007) Risk and vulnerability indicators at different scales: Applicability, usefulness and policy implications. Environmental Hazards 7: 20-31

Brooks, N. and Adger, W. N., 2003, Country level risk measures of climate-related natural disasters and implications for adaptation to climate change. Tyndall Centre Working Paper 26, as seen on 20th of January on http://www.tyndall.ac.uk/publications/working_papers/wp26.pdf

Brooks, N., Nicholls, R., and Hall, J. 2006. Sea Level Rise: Coastal Impacts and Responses, *Norwich WBGU* ISBN 3-936191-13-1 Berlin.

Brooks, N., W. N. Adger, and P. M. Kelly. The determinants of vulnerability and adaptive capacity at the national level and the implications for adaptation. Global Environmental Change, 15:151–163, 2005.

Brooks, Nick and W. Neil Adger. 2003. Country level risk measures of climate-related natural disasters and implications for adaptation to climate change. Tyndall Centre Working Paper 26. Norwich: Tyndall Centre for Climate Change Research. 26pp.

Brown, A.C., and A. McLachlan. 2002. Sandy shore ecosystems and the threats facing them: some predictions for the year 2025. *Environ. Conserv.*, 29, 62-77 Cambridge, pp. 289–324.

Brunsden, D. & Chandler, J.H. (1996) Development of an episodic landform change model based upon the Black Ven Mudslide, 1946–1995. Advances in Hillslope Processes: Volume 2, ed. M.G. Anderson & S.M. Brooks, pp. 869–896. Chichester, UK: John Wiley.

Bryan, B., Harvey, N., Belperio, T. and Bourman, B., 2001. Distributed process modeling for regional assessment of coastal vulnerability to sea-level rise. Environmental Modeling and Assessment, 6. 57–65

Burton, C., Cutter, S.L., 2008, Levee Failures and Social Vulnerability in the Sacramento-San Joaquin Delta Area, California, Natural Hazards Review, Vol. 9, No. 3, 136-149, August 1, 2008.

Canadian Council of Ministers of the Environment, 2003, Climate, Nature, People: Indicators of Canada's changing climate, as seen in www.ccme.ca , on 22nd of November, 2006

Cardona, O., 2003, A need for rethinking the concept of vulnerability and risk from a holistic perspective: A necessary review and criticism for effective risk management, as see on http://www.desenredando.org/public/articulos/2003/nrcvrfhp/nrcvrfhp_ago-04-2003.pdf ,on 8th of November, 2006

Carter, R.W.G. (1988) Coastal Environments: An Introduction to the Physical, Ecological and Cultural Systems of Coastlines. London, UK: Academic Press: 614 pp.

Centre for Research on the Epidemiology of Disasters (CRED) Université catholique de Louvain, as see on www.cred.be, on 29th of January 2010

Centre for Research on the Epidemiology of Disasters (CRED), 2009. Universite' catholique de Louvain. www.emdat.be/Documents/CredCrunch/ CredCrunch17.pdf (accessed 24 July 2009).

Centre for Research on the Epidemiology of Disasters (CRED), 2010. Universite' catholique de Louvain. http://cred.be/sites/default/files/CredCrunch19.pdf (accessed 26 February 2010).

Chambers, R. (1989), "Editorial introduction: vulnerability, coping and policy", IDS Bulletin, Vol. 20 No. 2, pp. 1-7.

Chander, R. (1996), Measurement of the Vulnerability of Small States Report prepared for the Commonwealth Secretariat, April 1996.

Chau, K.,W., 1990, Application of the Preissmann scheme on flood propagation in river systems in difficult terrain Hydrology in Mountainous Regions. I - Hydrological Measurements; the Water Cycle (Proceedings of two Lausanne Symposia, August 1990). IAHS Publ. no. 193

Chow V.T., Maidment, D. R., Mays, L.W. 1988 Applied Hydrology, McGraw-Hill, New York

Connor, R. F. and Hiroki, K., 2005. Development of a method for assessing flood vulnerability. Water Science and Technology, 51(5). 61–67.

Cooper, J.A.G., and F. Navas. 2004. Natural bathymetric change as a control on century-scale shoreline behaviour. *Geology*, 32, 513-516.

Cova T.J., Church R. L., 1997, Modelling community evacuation vulnerability using GIS Journal of geographical information science, vol. 11, no. 8, pg. 763± 784

Cowell, P.J., M.J.F. Stive, A.W. Niedoroda, H.J. De Vriend, D.J.P. Swift, G.M. Kaminsky andM. Capobianco. 2003a. The coastal tract. Part 1:Aconceptual approach to aggregatedmodelling of low-order coastal change. *J. Coastal Res.*, 19, 812-827.

Cowell, P.J.,M.J.F. Stive, A.W. Niedoroda, D.J.P. Swift, H.J. DeVriend,M.C. Buijsman, R.J. Nicholls, P.S. Roy and Co-authors. 2003b. The coastal tract. Part 2: Applications of aggregated modelling of lower-order coastal change. *J. Coastal Res.*, 19, 828-848.

CRED CRUNCH, 2010, Disaster data. A balanced perspective, issue 21, August 2010, Centre for Research on the Epidemiology and Disasters, Universite Catholique de Louvain as seen on 9th September 2010 on http://www.cred.be/sites/default/files/CredCrunch21.pdf

Crooks, S.. 2004. The effect of sea-level rise on coastal geomorphology. Ibis, 146, 18-20.

Crowards, T. (1999), An Economic Vulnerability Index , with Special Reference to the Caribbean: Alternative Methodologies and Provisional Results, Caribbean Development Bank, March 1999.

Cunge, J. A., Holly, F. M., Verwey, A., 1980, Practical Aspects of Computational River Hydraulics

Cutter S.L., 1996 Vulnerability to environmental hazards; Progress in Human Geography 20(4):529-539.

Cutter, S. L. B, J. Bi.rulT atid W. L. Shirley, "'Social Vulnerability to Environmental Hazards." Social Science Quarterly U. nu. 1 (2(X)3): 242-61; I^he Hein/ Center. Hwnim Links lo Coasusl Disasters (Washington. W: I'he H. J()bn Hein/ III Center for Seienee. Heiinomies and the Environment. 2(M)2). 57-114

Cutter, S. L., 2005, The Role of Vulnerability Science in Disaster Preparedness and Response Research. Testimony provided to the Subcommittee of the U.S. House of Representatives' Committee on Science, "The Role of Social Science Research in Disaster Preparedness and Response", November 10, 2005 Available from http://science.house.gov/publications/hearings_markups_details.aspx?NewsID=976, as seen on April 2009

Cutter, S. L., 2006, Hazards, Vulnerability, and Environmental Justice. London and Sterling, VA: Earthscan. 418 pp.

Cutter, S. L., Boruff, B. J. and Shirley,W. L., 2003. Social vulnerability to environmental hazards. Social Sciences Quarterly, 84(2). 242–261.

Cutter, S.L. (2005), "Are we asking the right questions?", in Perry, R.W. and Quarantelli, E.L. (Eds), What is a Disaster? New Answers to Old Questions, Xlibris, Philadelphia, PA, pp. 39-48.

Cutter, S.L., 1993, Living with risk. London, Edward Arnold

Cutter, Susan L. Christopher I Emnch, Jerry I Mitchell Bryan J. Boruff, Melanie Gall, Mathew C. Schmidtiein, Christopher G. Burton, and Ginni Melton., 2006, The long road home: Race, Class, and Recovery from Hurricane Katrina, Environment, volume 48, number 2

Dao, H., P., Peduzzi, (2004), Global evaluation of human risk and vulnerability to natural hazards, EnviroInfo 2004 conference proceeding, Geneva.

Dartmouth Flood Observatory, 2009, Active Archive of Large Floods, 1985-Present, as seen on: http://www.dartmouth.edu/~floods/, on November, 2010

Dasgupta, S., Laplante, B., Meisner, C., Wheeler, D., Jianping, Y. 2007. The impact of sea level rise on developing countries : a comparative analysis, Policy Research Working Paper, World Bank, WPS4136.

Davison C. M., Robinson S., Neufeld V., 2004, Mapping the Canadian Contribution to Strengthening National Health Research Systems in Low and Middle Income Countries: A concept paper, as see on http://www.ccghr.ca/docs/ConceptPaperPostWrkshp.doc on September 2010

Dawson, R. J., Dickson, M., Nicholls, R. J., Hall, J., Walkden, M. J. A., Stansby, P. K., Mokrech, M., Richards, J., Zhou, J., Milligan, J., Jordan, A., Pearson, S., Rees, J., Bates, P. D., Koukoulas, S. and Watkinson, A., 2009, Integrated analysis of risks of coastal flooding and cliff erosion under scenarios of long term change. Climatic Change, 95(1–2). 249–288

Day, 2005 Day, A.-L., 2005. Carlisle storms and associated flooding: multi-agency debrief report. Technical report, UK Resilience.

De Boer, G. (1988) History of the Humber coast. In: A Dynamic Estuary: Man, Nature and the Humber, ed. N.V. Jones, pp. 16 30. Hull, UK: Hull University Press.

De Bruijn, K.M., 2004. Resilience indicators for flood risk management systems of lowland rivers. International Journal of River BasinManagement, 2(3). 199–210.

De Leon, V., 2006. Vulnerability, a conceptual andmethodological review, UNU-EHS, SOURCE no4/2006. Measuring Vulnerability to Natural Hazards: Towards Disaster Resilient Societies, Chapter 3, J. Birkmann (ed.). United Nations University Press, New York.

De Waal, A., 1989. Famine That Kills: Darfur, Sudan, 1984–1985. Clarendon Press, Oxford.

Debasri, R., Sandip, M., Belaram, B. 1995. Man's Influence on Freshwater Ecosystems and Water Use, Proceedings of a Boulder Symposium IAHS no 230, pp. 95-100.

DeFries, R.S., Rudel, T., Uriarte, M. & Hansen, M., 2010. Deforestation driven by urban population growth and agricultural trade in the twenty-first century. Nature Geoscience, [Online] 3, Available at: http://www.nature.com/ngeo/journal/v3/n3/full/ngeo756.html [Accessed 16 January 2012].

Deltacommissie. 2008. Working together with water - A living land builds for its future – *Findings of the Deltacommissie* 2008, Rotterdam, 2008, pg. 117.

Dessai S, Hulme M. 2007. Assessing the robustness of adaptation decisions to climate change uncertainties: a case study of water resources management in the East of England. Global Environmental Change 17: 59–72.

Di Baldassarre, G., A. Castellarin, A. Montanari, A. Brath, 2009. Probability weighted hazard maps for comparing different flood risk management strategies: a case study, Natural Hazards, 50(3), 479-496.

Di Baldassarre, G., G. Schumann, P. Bates, J. Freer, K. Beven, 2010. Floodplain mapping: a critical discussion on deterministic and probabilistic approaches, Hydrological Sciences Journal, 55(3), 364-376.

Di Mauro, C. 2006. Regional vulnerability map for supporting policy definitions and implementations", ARMONIA Conference *"Multi-Hazards: Challenges for Risk Assessment, Mapping and Management"*, Barcelona.

Dinh, Q., Balica, S., Popescu, I., Jonoski, A., (2012), Climate change impact on flood hazard, vulnerability and risk of the Long Xuyen Quadrangle in the Mekong Delta, International Journal of River Basin Management, 10:1, 103-120

Doerfliger N, Zwahlen F (1998) Practical guide, groundwater vulnerability mapping in karstic regions (EPIK). Swiss Agency for the Environment, Forests and Landscape (SAEFL), Bern, 56 pp

Doerfliger N., Jeannin P-Y., Zwahlen F., 2000, Water vulnerability assessment in karst   environments: a new method of defining protection areas using a multi-attribute   approach and GIS tools (EPIK method), University of Neuchatel, from SAEFL, as seen on   http://www.acsad-bgr.org/files/gw_vul_annex3_epik.pdf on 5th of   February 2008

Dominguez, L., Anfuso, G. and Gracia, F. J., 2005, Vulnerability assessment of a retreating coast in SW Spain. Environmental Geology, 47. 1037–1044

Doracie B. Zoleta-Nantes. 1999. The Flood Landscapes of Metro Manila, Chronicle vol. 4 nos. 1-2.

Douben, K-J., 2006a, Characteristics of river floods and flooding: a global overview, 1985 – 2003, Irrigation and Drainage 55, S9 –S21, published online in Wiley InterScience

Douben, K-J., 2006b, Flood Management, UNESCO-IHE, Lecture Notes

Douben, N., Ratnayake, RMW, 2005, Characteristic data on river floods; facts and figures. In Floods, from Defense to Management, Symposium Papers, van Alphen J, van Beek E, Taal M (eds). Taylor & Francis Group: London, UK; pp 11-27

Doukakis, D., 2005, Coastal Vulnerability and Risk Parameters, European Water 11/12: 3-7. E.W. Publications

Downing, T.E. and Patwardhan, A., with Klein, R.J.T., Mukhala, E., Stephen, L., Winograd, M. and Ziervogel, G. 2005, Assessing Vulnerability for Climate Adaptation; In Adaptation Policy Frameworks for Climate Change: Developing Strategies, Policies and Measures. Lim, B., Spanger-Siegfried, E., Burton, I., Malone, E. and Huq, S. (Eds). Cambridge University Press, Cambridge

Downing, T.E., 1991, Vulnerability to hunger and coping with climate change in Africa. Global Environmental change 1, 365-80

Dyson, L.L., van Heerden, J., 2001, The heavy rainfall and floods over the north-eastern interior of South Africa during February 2000, S. Afr. J. Sci. 97 (3–4), pp. 80–86

Easter, Christopher D. (1998), Small States and Development: A Composite Index of Vulnerability, Small States: Economic Review and Basic Statistics, Commonwealth Secretariat, December 1998.

Easter, Christopher. 1999. Small states development: a Commonwealth Vulnerability Index. The Round Table 351 : 403-422.

EC-JRC. 2005. Climate change and European Water Dimension, *EU Report No. 21553*, Steven J. Eisenreich.

Edgeworth, F. Y. (1925). The plurality of index numbers. The Economic Journal, 35(139), 379– 388.

Ericson, J.P., Vorosmarty, C.J. Dingman, S.L. Ward L.G. and Meybeck, M. 2006. Effective sea-level rise and deltas: causes of change and human dimension implications. *Global Planet Change*, 50, 63-82.

ESPON Hazards project, 2004, The spatial effects and management of natural and   technological hazards in general and in relation to climate change. 3rd Interim Report,   March

ESPON, 2006, http://www.espon.eu as seen on 29 September 2010-09-09

European Parliament and Council of the European Communities. 2007. Directive on the assessment and management of flood risks (2007/60/EC). Official J L288/ 27–34.

Fekete, A., 2009, Validation of a social vulnerability index in context to river-floods in Germany Nat. Hazards Earth Syst. Sci., 9, 393–403, 2009 www.nat-hazards-earth-systsci.net/9/393/2009/

Fenster M.S., Dolan, R. 1996. Assessing the impact of tidal inlets on adjacent barrier island shorelines, *Journal of Coastal Research*, 12:294-310.

Fisher, I. (1922). The making of index numbers. Journal of the American Statistical Association.

Floods, 2005, Emergency and disasters, Management Inc., as see on http://www.emergencymanagement.net/flood.htm in October, 2009

Forbes, D.L., Orford, J.D., Carter, R.W.G., Shaw, J. & Jennings, S.C. (1995) Morphodynamic evolution, self-organisation, and instability of coarse clastic barriers on paraglacial coasts. Marine Geology 126: 63–85.

Gabor, T., Griffith, T.K., 1980, The assessment of community vulnerability to acute hazardous materials incidents, Journal of Hazardous Materials 8, 323-22

Galderisi, A., Ceudech, A., Pistucci, M., 2005, Integrated vulnerability assessment: the relevance "to" and "of" urban planning. In: Proceedings (CD format) of the 1st ARMONIA Project Conference "Multi-hazards: challenges for risk assessment, mapping and management", Barcelona, 5–6 December 2005

Gallopin, G. C., 2006, Linkages between vulnerability, resilience, and adaptive capacity. Glob. Environ. Change 16(3), pp. 293–303

Gaume, E., Gaál, L., Viglione, A., Szolgay, J., Kohnová, S., Blöschl G., 2010, Bayesian MCMC approach to regional flood frequency analyses involving extraordinary flood events at ungauged sites Journal of Hydrology, Volume 394, Issues 1-2, 17 November 2010, Pages 101-117

Geography website, 2009, as seen on www.geography.org.uk/, on March, 2009

Gheorghe A. (ed.) (2005). Integrated Risk and Vulnerability Management Assisted by Decision Support Systems. Relevance and Impact on Governance, Springer, Dordrecht as seen on http://www.caenz.com/info/2005Conf/pres/Gheorghe.pdf on 1st of December, 2006

Githui F. W., 2008, Assessing the impacts of environmental change on the hydrology of the Nzoia catchment, in the Lake Victoria Basin. Department of Hydrology and Hydraulic Engineering ,Faculty of Engineering Vrije Universiteit Brussel Pleinlaan 2 1050 Brussels

Githui F., Gitau W., Mutuab F., Bauwensa W., (2008) Climate change impact on SWAT simulated streamflow in western Kenya. International Journal of Climatology 29: 1823 - 1834

Goldscheider N, Klute M, Sturm S, Hötzl H (2000) The PI method: a GIS based approach to mapping groundwater vulnerability with special consideration of karst aquifers. Z Angew Geol 463:157–166

Gomez, G., 2001. Combating Desertification in Mediterranean Europe: Linking Science with Stakeholders, contract EVK2-CT-2001-00109. King's College, London. www.kcl.ac.uk/projects/desertlinks.

Gommes, R. du Guerny, J., Nachtergaele, F., Brinkman, R. 1997. Potential impacts of sea-level rise on populations and agriculture. *FAO*, Rome.

Gornitz, V. 1991. Global coastal hazards from future sea level rise. *Palaeogeography, Palaeoclimatology, Palaeoecology*, vol 89, pp 379-398.

Gornitz, V. 2001. Sea-level rise and coasts. In Rosenzweig, C. and W.D. Solecki, (eds.), "Climate Change and a Global City: An Assessment of the Metropolitan East Coast Region" (pp. 21 – 46) *Columbia Earth Institute*, New York, 210 pp.

Gornitz, V. and P. Kanciruk. 1989. Assessment of global coastal hazards from sea-level rise. Coastal Zone '89. pp. 1345-59. *In Proceedings of Sixth Symposium on Coastal and Ocean Management*. ASCE, Charleston, South Carolina, pp. 1345-1359.

Gornitz, V., 1990, Vulnerability of the East Coast, USA to future sea level rise. Journal of Coastal Research, 1(Special Issue No. 9). 201–237

Gornitz, V.M., White, T.W. and Cushman, R.M. 1991. Vulnerability of the US to future sea level rise, Coastal Zone '91, *Proceedings of the 7th Symposium on Coastal and Ocean Management, American Society of Civil Engineers*, pp. 1345-1359.

Green, C. (2004). The evaluation of vulnerability to flooding, Disaster Prevention and Management, volume 13, pp. 323-329, Emerald Group Publishing Limited

Greenbaum, N., Schwartz, U., Bergman, N., 2010, Extreme floods and short-term hydroclimatological fluctuations in the hyper-arid Dead Sea region, Israel Global and Planetary Change, Volume 70, Issues 1-4, February 2010, Pages 125-137

Guillaumont, P. (2004a). 'A Revised EVI'. UN Document CDP/2004/PLEN/16, 31/03/2004. New York: UN.

Guillaumont, P.,(2008). "An Economic Vulnerability Index: Its Design and Use for International Development Policy", WIDER Research Paper, UNUWIDER, Vol. November 2008/ 99.

Gunton, A. (1997) Upper foreshore and sea wall stability, Jersey, Channel Islands. Journal of Coastal Research 13: 813–821.

Haase D. (2003) Holocene floodplains and their distribution in urban areas functionality indicators for their retention potentials. Landscape and Urban Planning 66: 5-18

Hall J. W., Evans Edward P., Penning-Rowsell Edmund C., Sayers Paul B., Thorne Colin R., Saul Adrian J., 2003, Quantified scenarios analysis of drivers and impacts of changing flood risk in England and Wales: 2030–2100, Environmental Hazards 5, pp. 51–65

Harley, C.D.G., A.R.Hughes, K.M.Hultgren, B.G.Miner, C.J.B. Sorte, C.S. Thornber, L.F. Rodriguez, L. Tomanek and S.L. Williams. 2006. The impacts of climate change in coastal marine systems. *Ecol. Lett.*, 9, 228-241.

Heyman B.N., Davis C., and Krumpe P.F., 1991, An assessment of world wide disaster vulnerability; Disaster Management 4:3-36.

Hinkel, J., "Indicators of vulnerability and adaptive capacity": Towards a clarification of the science–policy interface. Global Environ. Change (2010), doi:10.1016/j.gloenvcha.2010.08.002

Hiroyasu Kawai, Noriaki Hashimoto, Kuniaki Matsuura. 2008. Estimation of extreme storm surge and its duration in Japanese Bays by using stochastical typhoon model, *ICCE*, September 2008.

Holling C.S., 1973, Resilience and stability of ecological systems, Annual Review of Ecological Systems 4: 1-23.

Hölting B, Haertle T, Hohberger KH, Nachtigall KH, Villinger E, Weinzierl W, Wrobel JP (1995) Konzept zur Ermittlung der Schutzfunktion der Grundwasserüberdeckung [Concept for the evaluation of the protective function of the unsaturated stratum above the water table]. Geol Jahrb C63:5–24

Hoozemans, F.M.J., Stive, M.J.F., Bijlsma, L. 1993. Global vulnerability assessment: vulnerability of coastal areas to sea-level rise, *American Shore and Beach Preservation Association*, ASCE.

Horritt, M.S., Bates, P.D, Fewtrell, T. J., Mason, D. C., Wilson, M. D., 2010. Modelling the hydraulics of the Carlisle 2005 flood event, Proceedings of the Institution of Civil Engineers, Water Management, 163, 273–281.

Horritt, M.S., Bates, P.D., 2001. Effects of spatial resolution on a raster based model of flood flow. Journal of Hydrology 253, 239–249.

Hötzl, H., Adams, B., Aldwell, R., Daly, D., Herlicska, H., Humer, G., De Ketelaere, D., Silva, M.L., Šindler, M., Tripet, J-P., 1995, Chap 4, Regulations. In: European Commission: COST Action 65, Hydro-geological Aspects of Groundwater Protection in Karst Areas, Final report of the action. Report EUR 16547 EN, pp 403–434

Hoyt, G.W., Langbein, W., 1955, Floods, Princeton, New Jersey, Princeton University Press

Hughes, P. and Brundrit, G. B., An index to assess South Africa's vulnerability to sea-level rise. South African Journal of Science, 88(June 1992). 308–311.

Hughes, P. and Brundrit, G.B. 1992. An index to assess South Africa's vulnerability to sea-level rise. South African Journal of Science, 88: 308-311IPCC, 2007, Climate Change, Impacts, adaptation and vulnerability. contribution of working group II to the fourth assessment report of the intergovernmental panel on climate change, in: M.L. Parry, O.F. Canziani, J.P. Palutikof, P.J. van der Linden, C.E. Hanson (Eds.), *Intergovernmental Panel on Climate Change* (IPCC), Cambridge University Press, New York, 2007.

Hy Dao, Pascal Peduzzi, 2004, Global evaluation of human risk and vulnerability to natural hazards, Enviroinfo 2004, Sharing, Editions du Tricorne, Genève, ISBN 282930275-3, vol. I, p.435-446.

International Strategy for Disaster Reduction (ISDR), 2004, http://www.unisdr.org/disaster-statistics/occurrence-type-disas.htm cited on January 4, cited on 2010

IPCC, (1992). Emission scenarios for the IPCC: an update, in Houghton, J.T., Callnader, B.A., Varney, S.K. (Eds), Climate Change 1992: The Supplementary Report to the IPCC Scientific Assessment, Cambridge University Press, Cambridge, pp.75-95

IPCC, (2001). Climate change 2001: The scientific basis. Cambridge, Cambridge University PressISDR (2004) Living with floods, UN guidelines offer decision-makers hope to reduce flood losses, World Water Day (22 March) the publication Guidelines for reducing flood losses was launched in Geneva, New York, Bangkok and Harare

IPCC, 1992, Intergovernmental Panel on Climate Change, First Assessment WMO/UNEP, Overview and policymakers summaries, June 1992

IPCC, 1996, Intergovernmental Panel on Climate Change, Second Assessment Climate Change

IPCC, 2001, Intergovernmental Panel on Climate Change, Climate change 2001: The scientific basis. Cambridge, Cambridge University Press

IPCC, 2007, Intergovernmental Panel on Climate Change, AR4, Fourth Assessment Report: Climate Change 2007

ISDR, 2004, Living with floods, UN guidelines offer decision-makers hope to reduce flood losses, World Water Day (22 March) the publication Guidelines for reducing flood losses was launched in Geneva, New York, Bangkok and Harare

Jolliffe, I.T.,2002.PrincipalComponentAnalysis.2ndedn, Springer, University of Aberdeen, Aberdeen, UK.

Jones, R. and Boer, R., 2003, Assessing current climate risks Adaptation Policy Framework: A Guide for Policies to Facilitate Adaptation to Climate Change, UNDP, in review, as see on http://www.undp.org/cc/apf-outline.htm), on 15th of November, 2006

Jonkman, S.N., Vrijling, J.K., (2008) Loss of life due to floods. Journal of Flood Risk Management, 1: 43–56

Jørgensen, S.E., 1992, Integration of ecosystem theories: a pattern. Kluwer Academic Publishers, Dordrecht, The Netherlands.

Jun Jian. 2008. Simulated future river discharges under IPCC SRES scenarios: Yangtze, Ganges, Brahmaputra, Blue Nile and Murray-Darling Georgia Institute of Technology, Atlanta, GA; and P. Webster, Land-Atmosphere Interactions, Part VI (*Joint between the 22nd Conference on Hydrology and the 20th Conference on Climate Variability and Change*).

Kadoya, M., Chikamori H., Ichioka, T. 1993. Some characteristics of heavy rainfalls in the Yamato river basin found by the principal component and cluster analyses, Extreme Hydrological Events: Precipitation, Floods and Droughts (*Proceedings of the Yokohama Symposium*, July 1993). IAHS Publ. no. 213.

Kaly, U., Briguglio, L., McLeod, H., Schmall, S., Pratt, C. and Pal, R., 1999, Proceedings of the Environmental Vulnerability Index (EVI) Think Tank 7-10 September 1999. SOPAC Technical Report

Kasperson JX, Kasperson RE, Turner BL, Hsieh W, Schiller A (2005) Vulnerability to global environmental change. In: Kasperson JX, Kasperson RE (eds) The social contours of risk. Earthscan, London, pp 245–285

Kelly, P.M., Adger, W.N., (2000). Theory and practice in assessing vulnerability to climate change and facilitating adaptation, Climate change vol. 47, pg. 325-352, Kluwer Academic Publishing

Kenyon, W., 2007, Evaluating flood risk management options in Scotland: A participant led multi-criteria approach., *Ecological Economics, 64, 70-81*

Klein R.J.T., Nicholls R.J., Thomalla F., 2003, Resilience to natural hazards: How useful is this concept? Environmental Hazards 5 (2003) 35–45

Klein, R., 2004, Vulnerability Indices – An academic perspective, Expert Meeting "Developing a Method for Addressing Vulnerability to Climate Change and Climate Change Impact Management: To Index or Not To Index?" Bonn, Germany, 26 January 2004

Klein, R.J.T., R.J. Nicholls, and N. Mimura (1999). Coastal adaptation to climate change: Can the IPCC Technical Guidelines be applied? Mitigation and Adaptation Strategies for Global Change, vol. 4, pg 51-64.

Klijn, F., Bruijn, K., Ölfert, A., Penning-Rowsell, Simm, J. Wallis, M. (2009) Integrated Flood Risk Analysis and Management Methodologies, Flood risk assessment and flood risk management; An introduction and guidance based on experiences and findings of FLOODsite (an EU-funded Integrated project). http://www.floodsite.net /html/partner_area/search_results3b.asp?docID=798

Kron, W., 2005, Water International, Volume 3, Issue 1, 2005, pages 58-68, Taylor & Francis Group

Kron, W., 2008, Flood insurance: from clients to global financial markets, Journal of Flood risk management, DOI:10.1111/j.1753-318X.2008.01015.x

Kumar, P. K. Dinesh. 2006. Potential Vulnerability Implications of Sea Level Rise for the Coastal Zones of Cochin, Southwest Coast of India, *Environmental Monitoring and Assessment* (2006) 123: 333–344.

Kusumastuti, D. I., Sivapalan, M., Struthers, I., Reynolds, D. A., 2008, Thresholds in the storm response of a lake chain system and the occurrence and magnitude of lake overflows: Implications for flood frequency Advances in Water Resources, Volume 31, Issue 12, December 2008, Pages 1651-1661

Leichenko, R. M. and K. L. O'Brien (2002). "The Dynamics of Rural Vulnerability to Global Change: The Case of Southern Africa." Mitigation and Adaptation Strategies for Global Change 7(1): 1-18.

Leont'yev, I.O. 2003. Modelling erosion of sedimentary coasts in the Western Russian Arctic. *Coast. Eng.*, 47, 413-429.

Lewis, J. (1999), Development in Disaster-prone Places, Intermediate Technology, London.

Luers, A.L., Lobell, D.B., Sklar, L.S., Addams, C.L., Matson, P.A., 2003, A method for quantifying vulnerability, applied to the agricultural system of the Yaqui Valley, Mexico. Global Environmental Change 13, 255–267.

Marcoux, A., 2000. Population and deforestation. [Online] Sustainable Development Department of the Food and Agriculture Organization of the United Nations. Available at: http://www.fao.org/sd/wpdirect/WPan0050.htm [Accessed 15 January 2012].

McCarthy, J.J., Canziani, O.F., Leary, N.A., Dokken, D.J., White, K.S. (Eds.), 2001. Climate Change 2001: Impacts, Adaptation and Vulnerability. Cambridge University Press, Cambridge.

McCarthy, J.J., Canziani, O.F., Leary, N.A., Dokken, D.J., White, K.S. (Eds.), 2001. Climate Change 2001: Impacts, Adaptation and Vulnerability. Cambridge University Press, Cambridge.

McEntire, D., Crocker, C., and Peters, E., 2010, An Addressing vulnerability through an integrative approach", International Journal of Disaster Resilience in the Built Environment, Vol 1 No. 1, pp. 50-64.

McEntire, D.A. (2008), "Rising disasters and their reversal: an identification of vulnerability and ways to reduce it", in Pinkowski, J. (Ed.), Disaster Management Handbook, CRC Press, Boca Raton, FL, pp. 19-36.

McFadden, L. (2001). Developing an integrated basis for coastal zone management with reference to the eastern seaboard of Northern Ireland. Unpublished PhD Thesis, Queen's University of Belfast, Belfast, UK.

McLaughlin, S., and Cooper, J. A. G., 2010, A multi-scale coastal vulnerability index: A tool for coastal managers? Environmental Hazards: Human and Policy Dimensions, Volume 9, Number 3, 2010, Publisher: Earthscan pp. 233-248(16)

Measuring Vulnerability to Natural Hazards – Towards Disaster Resilient Societies" which was published by UNU-Press in October 2006

Menoni, S. and Pergalani, F., 1996, An attempt to link risk assessment with land use planning: a recent experience in Italy, Disaster Prev. Manage., 5, 6, 1996.

Merwade, V., Cook, A., and Coonrod, J. (2008) GIS techniques for creating river terrain models for hydrodynamic modelling and flood inundation mapping. Environmental Modelling & Software 23: 1300-1311.

Merz, B., Thieken, A. H. and Gocht, M., 2007. Flood risk mapping at the local scale: concepts and challenges. Flood Risk Management in Europe, S. Begum, M. J. F. Stive and J. W. Hall (eds). 231–251.

Messer, F., Meyer, V., 2005, Flood damage, vulnerability and risk perception – challenges for flood damage research, UFZ-Umweltforschungszentrum, Leipzig

Messner, F. &Meyer, V. 2006 Flood damages, vulnerability and risk perception—challenges for flood damage research. In: Schanze, J., et al. (eds) Flood Risk Management: Hazards, Vulnerability and Mitigation Measures. Springer. pp. 149–167.

Metcalfe, S.E., Ellis, S., Horton, B.P., Innes, J.B., McArthur, J., Mitlehner, A., Parkes, A. Pethick, J.S., Rees, J., Ridgway, J., Rutherford, M.M., Shennan, I. & Tooley, M.J. (2000) The Holocene evolution of the Humber Estuary: reconstructing change in a dynamic environment. In: Holocene Land–Ocean Interactions and Environmental Change around the North Sea, ed. I.Shennan & J. Andrews, pp. 97–118. London, UK: Geological Society, Special Publication 166.

Milly, P.C.D., Wetherald, R.T., Dunne, K.A., Delworth, T.L., 2002, Increasing risk of great floods in a changing climate. Nature, 415, pg. 514-517

Mirza M. Monirul Qader, 2003, Climate change and extreme weather events: can developing countries adapt? Climate Policy 3, pp. 233-248

Mitchell, J. (2002). Urban disasters as indicators of global environmental change: assessing functional varieties of vulnerability, Symposium on Disaster Reduction and Global Environmental Change, Federal Foreign Office, Berlin, Germany, June 20-21.

Mokrech, M., Nicholls, R. J., Richards, J. A., Henriques, C., Holman, I. P. and Shackley, S., 2008, Regional impact assessment of flooding under future climate and socio-economic scenarios for East Anglia and North West England. Climatic Change, 90. 31–55

Montz, B.E., Gruntfest, E., 2002, Flash Flood mitigation: recommendations for research and applications, Environmental Hazards 4, pp. 15-22

Moon, V.G. & Healy, T. (1994) Mechanisms of coastal cliff retreat and hazard zone delineation in soft flysch deposits. Journal of Coastal Research 10: 663–680.

MORESC, 2010, The Met Office of Rainfall and Evaporation Calculation System, as seen on http://www.nwl.ac.uk/ih/nrfa/yb/yb99/figure1_evap.htm, on 21st of November 2010

Moris-Oswald, T. and Sinclair, A. J., 2005. Values and floodplain management: case studies from the Red river basin. Environmental Hazards, 6. 9–22.

MunichRe, 2007, Flooding and insurance, Munich Reinsurance Company

Mustafa Daanish, 2003, Reinforcing vulnerability? Disaster relief, recovery, and response to the 2001 flood in Rawalpindi, Pakistan, Environmental Hazards 5 , pp. 71–82

Næss Lars Otto, Bang Guri, Eriksen Siri, Vevatne Jonas, 2005, Institutional adaptation to climate change: Flood responses at the municipal level in Norway, Global Environmental Change 15, pp. 125-138

NASA Earth Observatory, 2009, Floods across the Western Sahel, as seen on: http://earthobservatory.nasa.gov/NaturalHazards/view.php?id=40127, on November, 2010

Natural Environmental, 2007, Coastal Flooding, as see on http://www.heritage.nf.ca/environment/c_flooding.html on the 10th of December, 2010

Neelz, S., Pender, G., Villanueva, I., Wilson, M., Wright, N.G., Bates, P., Mason, D., Whitlow, C., 2005, Using remotely sensed data to support flood modelling, Proceedings of the Institution of Civil Engineers, Water Management 159, pp 35-43

Neukum C, Hötzl H., 2007, Standardization of vulnerability maps. Environ Geol 51:689–694

Neukum C., Hötzl H., Himmelsbach T., 2008, Validation of vulnerability mapping methods   by field investigations and numerical modelling, Hydrogeology Journal, Earth and   Environmental Science , ISSN1431-2174, April 2007

Neukum C., Hötzl H., Himmelsbach T., 2008, Validation of vulnerability mapping methods   by field investigations and numerical modelling, Hydrogeology Journal, Earth and   Environmental Science , ISSN1431-2174, April 2007

Nicholls Robert J., 2004, Coastal flooding and wetland loss in the 21st century: changes under the SRES climate and socio-economics scenarios, Global Environmental Change 14, pp 69-86

Nicholls, R, J., Wong, P. P., Burkett, V. R., Codignotto, J. O., Hay, J. E., McLean, R. F., Ragoonaden, S. and Woodroffe, C. D. 2007. Coastal systems and low-lying areas. In Climate Change 2007: Impacts, Adaptation and Vulnerability. *Contribution of Working Group II to the Fourth Assessment Report of the Intergovernmental Panel on Climate Change*, Cambridge University Press, Cambridge, UK.

NYC Hazards, 2007, Coastal Flooding, as see on http://www.nyc.gov/portal/site/nycgov/ on 10th of December, 2010

OCDE. 2007. Selon l'OCDE, le changement climatique pourrait multiplier par trois la population exposée à un risque d'inondations côtières d'ici 2070, seen on www.oecd.org/env/cc on 26th November 2009.

OCHA, 2009, United Nations Office for the Coordination of Humanitarian Affairs, West Africa - Flood Affected Population - June to September 2009, as seen on:

http://www.reliefweb.int/rw/rwb.nsf/db900sid/HHOO-7WEPSW?OpenDocument,   on
        November, 2010

Onywere, S. M., Z. M. Getenga., W. Baraza, S. S. Mwakalila, C. K. Twesigye, and J. Nakiranda, (2007),
        Intensification of Agriculture as the Driving Force in the Degradation of Nzoia River Basin: the
        Challenges of Watershed Management. Publication of the Lake Abaya Research Symposium
        2007 (LARS 2007) on Catchment and Lakes Research. May 7-11th 2007–Arba Minch, Ethiopia.
        FWU Water Resources Publications Vol: 06
        http://www.unisiegen.de/fb10/fwu/ww/publikationen/volume0607/index.html.en?lang=en

Orford, J.D., Carter, R.W.G., Jennings, S.C. & Hinton, A.C., 1995, Processes and timescales by which a
        coastal gravel-dominated barrier responds geomorphologically to sea-level rise – Story Head
        Barrier, Nova Scotia. Earth Surface Process and Landforms 20: 21–37.

Orford, J.D., Cooper, J.A.G. & McKenna, J., 1999, Mesoscale temporal changes to foredunes at Inch Spit,
        south-west Ireland. Zeitschrift für Geomorphologie 43: 439–461.

Ouarda, T.B.M.J., Cunderlik, J.M., St-Hilaire, A., Barbet, M., Bruneau, P., Bobée, B., 2006, Data-based
        comparison   of   seasonality-based   regional   flood   frequency   methods
        Journal of Hydrology, Volume 330, Issues 1-2, 30 October 2006, Pages 329-339

Page,   D.L.,   2000,   Floods: A   Predictable   Disaster
        (http://www.sn.apc.org/wmail/issues/0103316/OTHERS86.html)

Partyka, M.L., M. Weber, M.S. Peterson & S.T. Ross. 2005. Environment-habitat-production linkages in
        tidal river estuaries: conceptualisation through GIS, remote sensing, and tools of rapid data
        acquisition: a year in review. Joint GERS/SWS Chapter meetings, 30 March-2 April, Pensacola
        Beach, FL.

Peduzzi, P., Dao, H., Herold, C., Rochette, D. and Sanahuja, H., 2001. Feasibility Study Report – on Global
        Risk and Vulnerability Index –Trends per Year (GRAVITY). United Nations Development
        Programme Emergency Response Division UNDP/ERD, Geneva.

Pelling, M. 2003. The vulnerability of cities; Natural disaster and social resilience, Earthscan Publications,
        UK & USA.

Pelling, M., Uitto J., 2001, Small island developing states: natural disaster vulnerability and   global
        change. Environmental Hazards 3 pg. 49-62

Pender, G., and Néelz, S. (2007) Use of computer models of flood inundation to facilitate communication
        in flood risk management. Environmental Hazards, 7: 106-114.

Penning-Rowsell, E. C. and Chatterton, J. B., 1977. The Benefits of Flood Alleviation: A Manual of
        Assessment Techniques (The Blue Manual). Gower Technical Press, Aldershot, UK.

Penning-Rowsell, E. C. and Wilson, T., 2006. Gauging the impact of natural hazards: the pattern and cost
        of emergency response during flood events. Transactions, Institute of British Geographers,
        31(2). 9–15.

Penning-Rowsell, E., Floyd, P., Ramsbottom, D. and Surendran, S., 2005. Estimating injury and loss of life
        in floods: a deterministic framework. Natural Hazards, 36(1–2). 43–64.

Pérez España H. & Arreguín Sánchez, F., 1999, Complexity related to behaviour of stability in modelled
        coastal zone ecosystems. Aquatic Ecosystem Health and Management 2 (1999), pp. 129–135.

Perrow, C. (2006), "Disasters ever more? Reducing US vulnerabilities", in Rodrı´guez, H., Quarantelli, E.L.
        and Dynes, R.R. (Eds), Handbook of Disaster Research, Springer, New York, NY, pp. 521-33.

Perry, C., 2000, Significant Floods in the United States During the 20th Century – USGS Measures a
        Century of Floods, Fact Sheet 024 00 as see on http://ks.water.usgs.gov on the 8th of December,
        2010

Perry, R.W. (2006), "What is a disaster", in Rodrı´guez, H., Quarantelli, E.L. and Dynes, R.R. (Eds),
        Handbook of Disaster Research, Springer, New York, NY, pp. 1-15.

Pethick, J. and Crooks, S., 2000, Development of a coastal vulnerability index: a geomorphological perspective. Environmental Conservation, 27. 359–367

Pethick, J. S. (1992) Salt marsh geomorphology. In: Salt marshes: Morphodynamics, Conservation and Engineering Significance, ed. J.R.L. Allen & K. Pye, pp. 41–62. Cambridge, UK: Cambridge University Press.

Pethick, J.S. (1996) The geomorphology of mudflats. In: Estuarine Shores: Evolution, Environment and Human Health, ed. K.F. Nordstrom & C.T. Roman, pp. 185–211. Chichester, UK: John Wiley.

Philippines: Factor Analysis of Water-related Disasters in the Philippines, Technical Note of PWRI No. 4067, ISSN 0386-5878 prepared by ICHARM

Pirazzoli, P.A., H. Regnauld and L. Lemasson. 2004. Changes in storminess and surges in western France during the last century. Mar. Geol., 210, 307-323.

Plate Erich J., 2002, Flood risk and flood management, Journal of Hydrology 267 pp. 2-11

Prakasa, R., Murty, B.S. 2005. Estimation of flood vulnerability index for delta, Geoscience and remote sensing symposium Vol. 5 Nr. 25-29 July Pag. 3611-3614.

Pratt, C., Kaly, U. and Mitchell, J., 2004. How to Use the Environmental Vulnerability Index, UNEP/SOPAC South Pacific Applied Geo-science Commission. Technical Report 383. www.vulnerabilityindex.net/Files/ EVI%20Manual.pdf (accessed 17 November 2009).

Pritchard DW. 1967. What is an estuary: physical viewpoint, In: Lauff GH (Ed.), Estuaries, American Association for the Advancement of Science, Washington DC, pp. 3-5.

Prudhomme, C., Jakob, D., Svensson, C., 2003, Uncertainty and climate change impact on the flood regime of small UK catchments Journal of Hydrology, Volume 277, Issues 1-2, 1 June 2003, Pages 1-23

Quarantelli, E.L. (2005), "A social science research agenda for the disasters of the 21st century: theoretical, methodological and empirical issues and their professional implementation", in Perry, R.W. and Quarantelli, E.L. (Eds), What is a Disaster? New Answers to Old Questions, Xlibris, Philadelphia, PA, pp. 325-96.

Ramade, F., 1989, Ecological Catastrophes. Futuribles 1, 63-78

Rao P., Murty B.S., 2005, Estimation of flood vulnerability index for delta, Geosciences and   remote sensing symposium, Vol. 5 Nr. 25-29 July Pag. 3611-3614

Regnauld, H., P.A. Pirazzoli, G. Morvan and M. Ruz. 2004. Impact of storms and evolution of the coastline in western France. Mar. Geol., 210, 325-337.

Richie, W. & Penland, S. (1990) Aeolian sand bodies of the south Lousianna coast. In: Coastal Dunes, ed. K.F.Nordstrom, N.P. Pusty & R.W.G. Carter, pp. 105–27. Chichester, UK: John Wiley.

Ross, D.A. 1995. Introduction to Oceanography. New York: Harper Collins College Publishers. ISBN 978-0673469380.

Saalmueller, J., n.d.. Flood Management: Why it Matters for Development and Adaptation Policy. [Online] Available at: http://es.scribd.com/doc/53885460/Flood-Management-Why-It-Matters-for-Development-and-Adaptation-Policy [Accessed 13 January 2012].

Sayers P., Hall J., Dawson R., Rosu C., Chatterton J., Deakin R., Risk assessment of flood and coastal defenses for strategic planning (rasp) – a high level methodology. Wallingford, as seen on http://www.raspproject.net/RASP_defra2002_Paper_Final.pdf, on 29th of September, 2008

Schmocker-Fackel, P., Naef F., 2010, More frequent flooding? Changes in flood frequency in Switzerland since 1850 Journal of Hydrology, Volume 381, Issues 1-2, 5 February 2010, Pages 1-8

Schultz, B., 2006, Flood management under rapid urbanisation and industrialisation in flood-prone areas: a need for serious consideration, Irrigation and Drainage, Supplement: Integrated Flood Management Volume 55, Issue Supplement 1, pages S3–S8, July 2006

Sharples, C. 2006. Indicative Mapping of Tasmanian Coastal Vulnerability to Climate Change and Sea-Level Rise: *Explanatory Report* (Second Edition), Department of Primary Industries and Water, Hobart, Tasmania, pp. i-iv, 1-173. ISBN 0-7246-6385-1 [Revision/New Edition].

Sharples, C. 2008. The Smartline - an effective coastal data mapping format, *Coast to Coast Summary Report: Conference Presentations*, August, Darwin, NT EJ (2008) [Conference Extract].

Shaw, J., Taylor, R.B., Forbes, D.L., Ruz, M.-H. and Solomon, S. (1998), Sensitivity of the coasts of Canada to sea-level rise. Bulletin of the Geological Survey of Canada, vol 505, pp 1-79.

Simon, S., 2005. www.childrensmercy.org/stats/definitions/ correlation.htm.

Small, C., Nicholls, R.J. 2003. A global analysis of human settlement in coastal zones *Journal of Coastal Research* 19, 584–599.

Smit, B., Wandel, J. 2006. Adaptation, adaptive capacity and vulnerability, *Global Environmental Change*, 16, pp. 282- 292.

Smith, K., 2004. Environmental Hazards: Assessing Risk and Reducing Disaster. Routledge, London.

Smith, K., Ward, R. 1998. Floods: Physical Processes and Human Impacts. Wiley, Chichester, 382pp

Smithers, J.C., Schulze, R.E., Pike, A., Jewitt, G.P.W., 2001, A hydrological perspective of the February 2000 floods: a case study in the Sabie River Catchment. Water SA 27 (3), pp. 25–31

Sorenson, J.H. and Sorensen, B.V. (2006), "Community processes: warning and evacuation", in Rodrı́guez, H., Quarantelli, E.L. and Dynes, R.R. (Eds), Handbook of Disaster Research, Springer, New York, NY, pp. 183-99.

Sullivan C. A., Meigh J. R., Giacomello A. M., Fediw T., Lawrence P., Samad M., Mlote S., Hutton C., Allan J. A., Schulze R. E., Dlamini D. J. M., Cosgrove W., Delli Priscoli J.,   Gleick P., Smout I., Cobbing J., Calow R., Hunt C., Hussain A., Acreman M. C., King J.,   Malomo S., Tate E.L., O'Regan D., Milner S. and Steyl I. (2003). The Water Poverty   Index: development   and application at the community scale. Natural Resources Forum,   27, 189-199.

Sullivan C.A, 2002, Calculating a Water Poverty Index, Wallingford, World Development 7, pp 1195-1210

Sullivan, C. A. and Meigh, J. R., 2005. Targeting attention on local vulnerabilities using an integrated index approach: the example of the Climate Vulnerability Index. Water Science and Technology, 51(5). 61–67.

Sullivan, C.A, Meigh, J., 2003, Using the Climate Vulnerability Index to assess vulnerability to climate variations,   Water   Policy   and   Management,   CEH   Wallingford   as   seen   on: http://www.ceh.ac.uk/sections/ph/ClimateVulnerabilityIndex.html on 12th of December, 2006

Sultana, F., 2010, Living in Hazardous Waterscapes: Gendered Vulnerabilities and Experiences of Floods and Disasters Environmental Hazards, Bangladesh's Flood Action Plan at Twenty: Legacy of a Megaproject. Earthscan 43–53(11).

Tapsell, S. M., Penning-Rowsell, E. C., Tunstall, S. M. and Wilson, T., 2002. Vulnerability to flooding: health and social dimensions. Philosophical Transactions of the Royal Society of London, 360. 1511–1525.

Thieler, E.R. and Hammer-Klose, E.S. 2000. National Assessment of Coastal Vulnerabilityto Sea-Level Rise: Preliminary Results for the US Pacific Coast. Woods Hole, MA: United States Geological Survey (USGS), *Open File Report 00-178,* 1p.

Timmerman, P., 1981. Vulnerability, Resilience and the Collapse of Society: A Review of Models and Possible Climatic Applications. Institute for Environmental Studies, University of Toronto, Canada

Turner II, B.L., Kasperson, R.E., Matson, P.A., McCarthy, J.J., Corell, R.W., Christensen, L., Eckley, N., Kasperson, J.X., Luers, A., Martello, M.L., Polsky, C., Pulsipher, A., Schiller, A. (2003). A framework for vulnerability analysis in sustainability science. Proceedings of the National Academy of Sciences of the United States of America 100, 8074–8079.

Turner II., Matson B.L., McCarthy P.A., Corell J.J, Christensen R.W., Eckley L., Hovelsrud- Broda   N., Kasperson G.K., Kasperson J.X., Luers R.E., Martello A., Mathiesen M.,   Naylor   S.,   Polsky   R., Pulsipher C., Schiller A., Selin A., Tyler H., 2003b, Illustrating   the   coupled   human-environment system for vulnerability analysis: three case   studies.   Proceedings of the National Academy of Sciences US 100, 8080–8085

Turner, B. L., II et al, 2003, A framework for vulnerability analysis in sustainability science, as see on http://www.pnas.org/cgi/content/abstract/100/14/8074 on September 2010

UN/ISDR (United Nations International Strategy for Disaster Reduction). 2004. Living with Risk. A Global Review of Disaster Reduction Initiatives, 2004 version. United Nations: Geneva, 430 pp.

UN/ISDR, 2002 "Living with Risk: A Global Review of Disaster Reduction Initiatives" http://www.helid.desastres.net/

UNDP/BCPR, 2004, United Nations Development Programme and Bureau for Crisis Prevention and Recovery: A Global Report: Reducing Disaster Risk, A Challenge for Development, New York, 146 pp., ISBN 92-1-126160-0, Printed by John S. Swift Co. USA

UNDRO (United Nations Disater Relief Coordinator) (1979): Natural Disasters and Vulnerability Analysis in Report of Expert Group Meeting (9-12 July 1979), UNDRO, Geneva.

UNEP, 2004, Manual: How to Use the Environmental Vulnerability Index (EVI), as see on http://www.vulnerabilityindex.net/EVI_Calculator.htm , September 2010

UNEP. 2002. Global Environment Outlook 3, Earthscan, London, UK, pp. 446.

United Nation, UN (2003): 'Guidelines on participatory planning and management for flood mitigation and preparedness', ix, 129 s.: ill., Water resources series, 0082-8130, no. 82, United Nations: New York

United Nations Population Fund (UNFPA), 2011a. Linking Population, Poverty and Development (Rapid Growth   in   Less   Developed   Regions).   [Online]   Available   at: http://www.unfpa.org/pds/trends.htm [Accessed 13 January 2012].

United Nations Population Fund (UNFPA), 2011b. Linking Population, Poverty and Development (Urbanization: A   majority   in   cities).   [Online]   Available   at: http://www.unfpa.org/pds/urbanization.htm [Accessed 13 January 2012].

United Nations, 1982, Proceedings of the seminars on flood vulnerability analysis and on the principles of floodplain management for flood loss prevention, September, Bangkok

United States Agency, 1991, Department of Regional Development and Environment Executive Secretariat for Economic and Social Affairs Organization of American States With support from the Office of Foreign Disaster Assistance United States Agency for International Development Washington, D.C., Primer on Natural Hazard Management in Integrated Regional Development Planning

UNU-EHS, 2006, Vulnerability Indicators Floods Vulnerability Assessment of Rural and Urban Communities to Floods, as see on http://www.ehs.unu.edu , on 13th of November, 2006

USAID/FEWS Project: 1994, Vulnerability Assessment, Published for USAID, Bureau for Africa, Disaster response coordination, USAID/FEWS, Arlington, VA.

van Beek, E. & Loucks, D. P., 2005, Water Resources Systems Planning and Management—an introduction to methods, models and applications, UNESCO—Paris

van Beek, E., (2006) Water Resources Development, UNESCO-IHE, Lecture Notes

van der Sande, C.J., de Jong, S.M., de Roo A.P.J., 2003. A segmentation and classification approach of IKONOS-2 imagery for land cover mapping to assist flood risk and flood damage assessment. International Journal of Applied Earth Observation and Geoinformation 4 217–229.

van der Veen, A. & Logtmeijer, C. 2005 Economic hotspots: visualizing vulnerability to flooding. Nat. Hazards 36(1–2), 65–80.

Vaz, A.C., 2000, Coping with floods—the experience of Mozambique. In: Paper Presented in the International Conference on Mozambique Floods, Maputo, 27–28 October 2000, 15

Verwey, A., 2006, Computation Hydraulics, WL Delft Hydraulics, version 2006 – 1, UNESCO-IHE Lecture Notes

Villagran de Leon, J. C., 2006, Vulnerability - A conceptual and Methodological review, UNU EHS, no 4/2006, Bonn, Germany

Villarini, G., Smith, J.A., Serinaldi, F., Bales, J., Bates, P.D., Krajewski, W. F.,2009, Flood frequency analysis for nonstationary annual peak records in an urban drainage basin Advances in Water Resources, Volume 32, Issue 8, August 2009, Pages 1255-1266

Vincent, K., 2004, Creating an index of social vulnerability to climate change for Africa, Tyndall Centre for Climate Change Research, Working Paper 56

Walker, B., C. S. Holling, S. R. Carpenter, and A. Kinzig 2004, Resilience, adaptability and transformability in   social–ecological   systems.   Ecology   and   Society   9(2):   5, http://www.ecologyandsociety.org/vol9/iss2/art5/

Walstra, Dirk-Jan. 2009. Climate proofing coastal defence strategies, *Deltares* – Delft Hydraulics Delft University of Technology.

Ward, R., 1978. Floods, A Geographical Perspective. Macmillan Press, London.

Watson, R. T., Zinyowera, M. C. & Moss, R. H. (eds) 1996 'Climate Change 1995', in Impacts, Adaptations and Mitigation of Climate Change: Scientific-Technical Analyses. Cambridge University Press, Cambridge.

Watts M.J. and Bohle H.G., 1993, The space of vulnerability: the causal structure of hunger and famine. Progress in Human Geography 17:43-67.

Webster, P. J., Holland, G. J., Curry J. A., Chang H.R. 2005. Changes in Tropical Cyclone. Number, Duration, and Intensity in a Warming Environment ; *Science* Vol.309 no. 5742, pp. 1844 – 1846.

Wells, J. (1997), Composite Vulnerability Index: A Revised Report, Commonwealth Secretariat, London.

Wheater, H.S., 2006. Flood hazard and management: a UK perspective. Transactions of the RoyalSociety (http://rsta.royalsocietypublishing.org/content/364/1845/2135.full).

White, G.F., 1945. Human Adjustments to Floods: A Geographical Approach to the Flood Problem in the United States Doctoral Dissertation and Research paper no. 29. Department of Geography, University of Chicago.

White, P., Pelling, M., Sen, K., Seddon, D., Russell, S. and Few, R., 2005. Disaster Risk Reduction: A Development Concern. DFID, London.

Wilhelmi, O.V. and Wilhite, D.A., 2002, Assessing Vulnerability to Agricultural Drought: A Nebraska Case Study Natural Hazards 25: 37–58, 2002, Kluwer Academic Publishers. Printed in the Netherlands.

Wisner, B., Blaikie, P., Cannon, T. and Davis, I. (2004), At Risk: Natural Hazards People's Vulnerability, and Disaster, Routledge, London.

WMO/MWRMD/APFM, 2004, Strategy for Flood Management For Lake Vitoria Basin, Kenya. Associate programme on flood Management, http://www.apfm.info/pdf/strategy_kenya_e.pdf

Woodroffe, C.D. 2003. Coasts: Form, Process and Evolution. Cambridge University Press, 623 pp.

World Bank, 1994. Social Indicators of Development. World Bank, Washington, DC.

World Bank, 1997. Expanding the Measure of Wealth Indicators of Environmentally Sustainable Development, Rio 5th edn, discussion draft. World Bank, Washington, DC.

World   Bank,   2005.   African   Development   Indicators   2005. www4.worldbank.org/afr/stats/adi2005/adi05_booklet_   rev_061505.pdf   (accessed   16 November 2009).

World Meteorological Organization / Global Water Partnership (2007) Formulating a basin flood management plan A tool for integrated flood management. Associated programme on flood management.

Wright, N. G., Villanueva, I., Bates, P. D., Mason D. C., Wilson, M. D., Pender, G., Neelz. S., 2008, Case Study of the Use of Remotely Sensed Data for Modeling Flood Inundation on the River Severn, U.K., Journal of Hydraulic Engineering, ASCE May 2008, pp. 533-540

WWF, 2009. Mega-Stress for Mega-Cities A Climate Vulnerability Ranking of Major Coastal Cities in Asia, *WWF for a living planet report.*

Yohe, G., Malone, E., Brenkert, A., Schlesinger, M., Meij, Henk and Xing, Xiaoshi, 2006. Global Distributions of Vulnerability to Climate Change. Integrated Assessment, 6(3), 2006. ISSN 1389-5176. URL http://journals.sfu.ca/int_assess/index.php/iaj/article/view/239/210.

Yusuf, A.A. and H.A. Francisco. 2009. Climate Change Vulnerability Mapping for Southeast Asia, Economy and Environment Program for Southeast Asia (EEPSEA), Singapore www.eepsea.org

ACKNOWLEDGMENTS

The best and worst moments of my doctoral dissertation journey have been shared with many people. It has been a great privilege to spend several years in the Department of Hydraulic Engineering and River Basin Development, and its members will always remain dear to me. UNESCO-IHE is a place where I constantly feel inspired by the intelligence, culture and humanity surrounding me.

First and foremost, I would like to express my deep and sincere gratitude to my promoter and supervisor Prof. Nigel G. Wright, without his insightful supervision, invaluable guidance and humor from very early stages of this research, this thesis would not have been possible. As well, I would like to thank to my co-promoter, Prof. Arthur E. Mynett, for the unfailing support, a wise and caring advisor and simply a very good man in all the ways that truly matter.

I owe a great deal of appreciation for this thesis to Klass J. Douben, ir., as I like to say he was the "father" of this beautiful topic and he gave me the impulse in working consciously, with attitude. I thank him for his suggestions, refinements and constructive comments whenever I need it. Likewise, I am deeply grateful to Dr. Lindsay Beevers for her continuous encouragement and support, substantive guidance, valuable comments during the thesis.

As well, I would like to express my gratitude to the thesis committee members for their interest and valuable comments on my work.

I would like to thank to all academic staff for creating such an enjoyable and unique environment.

Generous financial support made my doctoral studies possible. I would like to thank the Netherlands Organisation for International Cooperation in Higher Education, Huygens Scholarship Programme for this privilege and the Rijkswaterstaat. This study has benefited from the administrative help of many people from UNESCO-IHE, I would also like to thank them all.

I have had the good fortune to cross paths with many dedicated teachers. Dr. Ioana Popescu, is happened to be a generous one, who gave and still gives me unconditionally knowledge about computational hydraulics and some other, on the other side she is one of my best friends, always sourced me a quality conversation about books, art and music, which made my work much poetical and inspired. Multumesc, Ioana!
Professor A. Lejeune, so dedicated to his work and trying all the ways which you can imagine to make us understand "Hydraulics", I just simply love him for his simplicity in teaching us. He was the one, without knowing, which gave me the wings to continue my studies and he cheered me up with his good attitude.

I have been very blessed throughout my life with quality friends. In all of the ups and downs that came my way during the "doctoral years", I knew that I had the support of several friends, be they near or far at any particular moment. Felipe Gonzales gave me sound advice, much-needed reassurance, and much joyful laughter, the gifts of a friendship that I believe will continue to stand the test of time and distance. Ali Dastgheib, a dear friend for many years, knows the best and worst of me, and continues to stick by me anyway. Ali, thank you for the countless ways in which you have supported me during this and other endeavors, and for being someone who just knew. You have my unending admiration and affection. Assiyeh Tabatabai, although we are not related by blood, I will always think of you as immediate family. Life has been made infinitely richer because of you, and I will spend the rest of my life trying to be the kind of friend to you that you have been to me. Mariana Popescu is simply unforgettable; I will remember the optimism she brings to everything she touches. Shilp Verma, you have been everything one could ask for in a friend. To Carlos and Carmen Lopez, I owe more than words can ever fully express. Thank you both for being with me to persevere during the bad times and celebrate the good times. Guy Beaujot, what life can be without a wise "old" man?

I wish to thank my parents, Alexandra Balica and Horia Balica. They born me, raised me, supported me, taught me, and loved me. To them I dedicate this thesis. Mami, Tati, sărut mâna și vreau să vă spun că sunteți înainte de toate lumina mea! Vreau să vă mulțumesc enorm pentru susținerea nelimitată morală și financiară, pentru sfaturile date în genere dar mai ales pentru acelea de a îndrăzni să cer mai mult și să ajung la stele și de a fi totuși cu picioarele pe pământ. Vă iubesc nelimitat! Also I wish to thank my sister, Daniela, and my brother in law, Cristian, for their support, love, good disposition and for teaching me that "nothing in life is impossible, so go for it and have no regrets". Is solace anywhere more comforting than in the arms of a sister?

Finally, this dissertation is dedicated also to my greatest blessing, my husband Sergiu Chelcea, the most decent and loving person I've ever known. In spite of your desire to pass quietly through life, unnoticed, all of us who have you in our lives know just how amazing you are in everything that you do. Simply watching you go about your daily life makes me proud to be your wife. Thank you for all your unconditional love and understanding. A lifetime with you will always be too short.

Curriculum Vitae

Stefania - Florina Balica, Romanian, born on 23rd of April, 1978, in Drobeta Turnu Severin, has been engaged in research in the field of flood vulnerability for nearly 10 years. Her first contact with the research was during her first year of fieldwork at the Department of Hydro-technical Engineering at Polytechnic University of Timisoara (Romania). Then she continued with the research for three years when she worked on a project on optimising diversion schemes using non-structural measures (Banat Watershed, Romania).

Following this she won a scholarship to study MSc at UNESCO-IHE, Delft, the Netherlands in Hydraulic Engineering and River Basin Development. There she have studied river dynamics, water resources systems planning and management, flood management, and finalised her degree with a thesis which developed a method to assess flood vulnerability in various spatial scales. After the Masters she returned back to Romania to the same University and worked as a researcher in the Department of Hydro-technical Constructions to continue her research, for a period of six months. At this point she received a prestigious award (Huygens Scholarship Programme) to continue her research at UNESCO-IHE and Delft University of Technology as a PhD student. Her PhD thesis focused on the application of the Flood Vulnerability Index methodology as a knowledge base for flood risk assessment, focusing on systems components, vulnerability factors and indicators. The methodology was initially developed for river flooding, but it has since been extended to coastal flooding at various spatial scales. The results of her PhD research are presented in this thesis.

Journal Publications:

S.F., Balica, Wright, N.G., van der Meulen, F., 2012, **A flood vulnerability index for coastal cities and its use in assessing climate change impacts**, *Natural Hazards*, Springer Publisher, accepted on 17th May 2012

S.F., Balica, 2012, **Approaches of understanding vulnerability indices developments to natural disasters**, *Environmental Engineering and Management Journal*, June, 2012, Vol. 12 (6)

Q., Dinh, S.F., Balica, I., Popescu, Jonoski, A., (2012), **Climate change impact on flood hazard, vulnerability and risk of the Long Xuyen Quadrangle in the Mekong Delta**, *International Journal of River Basin Management*, 10:1, 103-120

S.F., Balica, N.G., Wright, (2010). **Reducing the complexity of Flood Vulnerability Index**, *Environmental Hazard Journal*, EHJ, 9 (4). 321 - 339. ISSN 1747-7891

S.F., Balica, N., Douben, N.G., Wright, (2009). **Flood Vulnerability Indices at Varying Spatial Scales**, *Water Science and Technology Journal*, WST 60.10. 2009, pg. 2571-2580

S.F., Balica, N.G., Wright, (2009). **A network of knowledge on applying an indicator-based methodology for minimizing flood vulnerability**, *Hydrological Processes Journal*. Volume 23 (20), pages 2983-2986, 25th of August 2009

Conference Publications:

S.F., Balica, (2007) **Conceptualizing and visualizing flood vulnerability**, Workshop, Integrate Water Management, The Third Edition, 30-31.05.2007, Editura Orizonturi Universitare, ISBN: 978-973-638-326-7, Timisoara, Romania

S.F., Balica and F. Gonzalez Pallais, (2006) **GIS in policy making**, Preventing and fighting hydrological disasters 2, July 2006, Timisoara, Romania

S.F., Balica and G. Balla. (2006) **Forecast and Public Protection in the Banat Watershed using ArcGIS** 9.1, Preventing and fighting hydrological disasters 2, July 2006, Timisoara, Romania

G., Cretu, C., Rosu, F., Mocanu, S.F., Balica, A., Pepa, A., Riti, **Above the Floods Occurred in April-May 2005, in the Banat Catchment's Area**, November 2005, Politehnica, Iasi

G. Balla and S.F. Balica. **Sediment Transport in Watershed**, In Scientific Bulletin of "Politehnica" University of Timisoara, 13 May 2005, Timisoara, Romania

G. Balla and S.F.Balica, **The Importance of Water Resources Management**. In Water for Life, The Sustainable Development of Water Resources, 22 March 2005, Timisoara, Romania.

S.F. Balica **Details of Renaturation Techniques of Some Water Bodies**, Integrated Water Resources Management, May 2005, Timisoara, Romania

S.F. Balica and G. Balla, **Watershed's Arrangement Schemes at High Water**, In World Water Day, Water and Disasters, 22 March 2004, Timisoara, Romania

T - #0107 - 071024 - C170 - 244/170/9 - PB - 9780415641579 - Gloss Lamination